INGENUITY FOR A VITAL INDUSTRY

Few products produced by man have more utility and versatility than paper. Our laws, history, and literature are recorded on paper. Basic necessities are packaged in it. Mass communications are transmitted on it. The applications are infinite.

Due to recent ideas and innovations in the pulp and paper industry, the competition from industries such as plastics and electronics has been met. Although remarkable progress has been made, the research and development work continues to achieve new results in making better paper products with lower costs.

TVW — Dedicated to the Future of Pulp and Paper. The TVW Group is a joint marketing venture established by three major Finnish corporations: Tampella, Valmet, and Wärtsilä.

These three companies have served the pulp and paper industry in Scandinavia for nearly a century, and have provided pulp and papermaking technology throughout the world for the past fifty years.

The TVW Group is dedicated to ensuring that paper remains a viable, competitive product, and that pulp and papermaking becomes an even more integral part of the world economy in the future. The Group is committed to providing the industry with the most advanced technology possible and can show real achievements in overcoming the problems which the industry faces: high energy and labor costs, shortages of raw material, and environmental requirements. TVW will persist in seeking solutions to the industry's challenges in the future.

Today and tomorrow, TVW will continue to uphold its uncompromising tradition: INGENUITY FOR A VITAL INDUSTRY.

TVW

TAMPELLA • VALMET • WÄRTSILÄ

PAPER MACHINE GROUP

PAPERMAKING

The art and craft of handmade paper

By Ralf Weidenmüller

Translated from the German by John Kalish

Published by Thorfinn International Marketing Consultants Inc.
San Diego and Ralf E. Soderholm, London

Originally published in 1980 under the title:
Papiermachen: ein neues Hobby

Exclusive North American Distributors:
Miller Freeman Publications, Inc.
500 Howard Street
San Francisco, California 94105

ISBN: 0-87930-157-0
Layout and cover design: A. Poyblant

Contents

Translator's foreword

Paper is an unusual product. Perhaps it is as unique as papermakers like to claim. Certainly, it is one of the few products which pre-date the industrial revolution for which demand and consumption, worldwide, are still rising. Though paper is made today on very sophisticated equipment, the basic process itself differs little from that employed centuries ago. And while a single modern paper machine might be able to produce in a few days or weeks the equivalent of the entire world's production of 1000 years ago, the product itself is remarkably similar to its predecessor and would certainly be recognized by an ancient papermaker.

Many books have been written about paper and papermaking. This book, however, is unique, I believe. In a very compact and readable format it covers paper's history from discovery to the modern era; it provides a first class course in hand papermaking with a thoroughness and detail not found elsewhere; and it offers the reader a concise summary of modern industrial papermaking. And all this in only about 60 pages. With this book, virtually anyone can grasp the fundamentals of paper and papermaking in a few hours, whether or not one has any particular educational background for the subject. This book can be read simply for general knowledge or out of curiosity, or it can serve as a fully detailed instruction book for hand papermaking, up to and including the crafting of difficult watermarks and building one's own papermaking equipment. *Papermaking* is just as readable for a young person of 12 or 13 as it is for his or her parents, or grandparents for that matter. *Papermaking* should prove interesting for the casual reader; challenging for the ambitious hand papermaker.

In translating this book I have departed from papermakers' jargon in only one area: instead of forming mold, I have called it a forming frame, since that is what it looks like to most people who see it for the first time. Similarly, I have not used the term "deckle". The present text reflects numerous corrections and amendments by the author, and the section on modern papermaking has been updated to take account of developments made during the five years since the original German text was written.

A special word of appreciation is offered to co-publisher Ralf Soderholm for his careful proofreading and correction of the text, and to my colleague Leonard Haas for his thoughtful observations and advice during this book's production.

Brussels, August 1984

PAPER IN OUR WORLD

Paper is taken for granted throughout the civilized world. Yet it would be hard to imagine our daily lives today without paper — whether for written or printed information, for magazines or books, for packaging or for hygienic and household uses.

In addition to the commonly known papers, there are many beautiful and valuable papers, which, although they are machine-made today, have come down to us from the times of hand papermaking. And this book is about making paper by hand. This fine and honored craft was pushed aside by the rapid industrialization of the past 100 years, and has been almost forgotten in many places.

In centuries past, papermaking had grown to become a prized craft. People spoke of "the art of papermaking". And indeed, anyone who has seen a sheet of handmade paper being formed cannot help but be fascinated.

With its links to the history of our culture and civilization, to literature and knowledge, papermaking attracts the talented craftsman and the intellectual alike. Papermaking offers satisfaction to everyone — children, adults, and entire families. In fact, papermaking is one of the few crafts which can give pleasure throughout a lifetime, providing not just personal satisfaction but many useful and beautiful papers for your own use as well as for very attractive gifts.

This book give you not only a complete and precise description of the hand papermaking process, but also a concise perspective into the methods of ancient and modern papermaking.

FIGURE 1: Ulmann Stromers paper mill near the gates of Nürnberg, from the Schedel'sche Chronical of 1493.

Chapter I

THE HISTORY OF PAPER

Throughout the history of mankind, the availability of a writing material has always promoted the development of the written language.

Primitive, pre-historic peoples developed knowledge which they wanted to record; thus, understandably, the first written symbols known to us were found in the drawings on the walls of caves.

To transmit these recorded ideas from one place to another, some transportable medium had to be found. The first was the clay tablet: the writing was done with a stylus in the wet clay, then the tablet was either dried in the sun or baked in a fire. Later, sheets of soft metal such as lead or copper were used in similar fashion.

The Greeks and Romans used wooden tablets coated with wax. Leaves of plants were also used as writing materials, and it is known that the bark of the white birch tree served this purpose in more recent times due to lack of paper.

The idea to imitate the bark of trees and make larger pieces of a writing material was probably what brought the Egyptians to the discovery and production of *Papyrus*. The making of this writing material is said to have been first achieved in the 4th century before Christ.

How papyrus sheets were made

Papyrus is a plant native to the marshland of tropical Africa. Its three-sided stalks grow to a height of about 5 meters (16 feet) and are 6 to 8 centimetres (2—3 inches) thick at the base.

To make papyrus sheets, the first step was to cut the fairly tough inner core of the plant's stem with a knife into long, thin and preferably wide strips. These strips were then laid next to each other on a board, with edges slightly overlapping, and then a second layer was laid in the same fashion perpendicular to the first.

The sheet made of these two layers of papyrus strips was then pressed and beaten with a hammer so that its thickness was uniform. The sap squeezed out of the strips by this hammering served to bond the fibers together better. The sheets were then smoothed with a shell or an animal tooth. The finished sheets were light yellow in color, thin, smooth and durable.

FIGURE 2: Papyrus plants. Below, section through a papyrus stalk, and how the strips of papyrus were put together to make a sheet.

Papyrus scrolls and boxes have held up well over thousands of years in the dry climate of the Mediterranean, and have been found in recent times in good condition. In Egypt, for example, a great number of papyrus scripts have been preserved.

Egypt, in its declining years, supplied the far-flung Roman Empire with its papyrus. Up until the Middle Ages, the royal paper mills of Egypt continued to make papyrus.

The relatively small papyrus sheets — a maximum of 50 cm long (20 inches) — could not be made into books because they could not be easily folded. Instead, papyrus sheets were often glued together end-to-end to make long sheets — up to 20 meters (65 feet) long — and then rolled into scrolls.

However, although the word paper is derived from papyrus, this ancient writing material is not really a form of paper by modern definition. As it is built up of layers, papyrus is really more similar in structure to plywood.

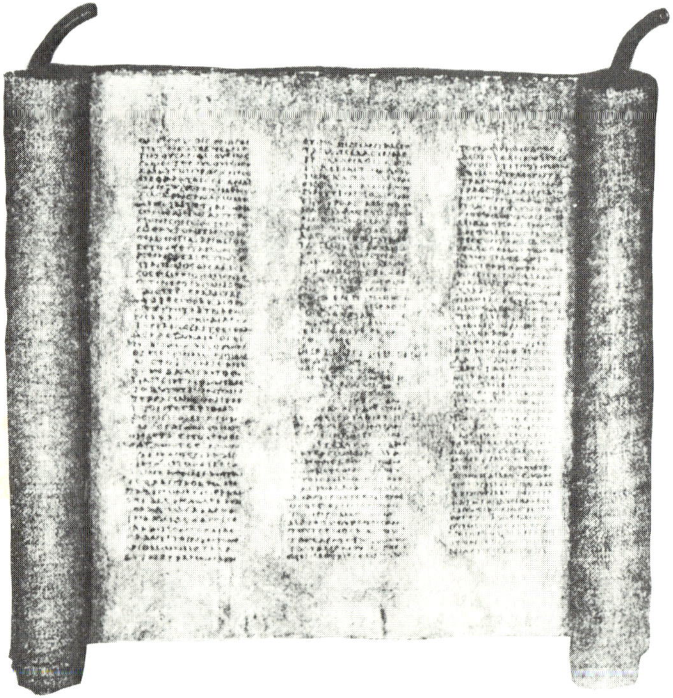

FIGURE 3: A papyrus scroll.

Animal skins as writing material

Another writing material, even older than papyrus, is made from the untanned skins of young sheep and goats. This writing material is most likely older than the Roman historian Pliny reports. He placed its introduction in the 2nd century before Christ in the city of Pergamon in Asia Minor. And today, this writing material is still called *Pergament.*

Pergament processing came from Italy to Germany

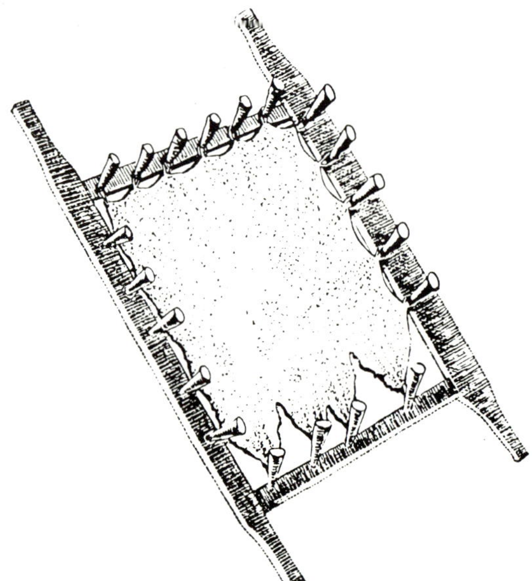

FIGURE 4: Pergament was made by stretching animal skins over a wooden frame, as shown here.

and France and through the rest of Europe. The process comprises softening the animal skins, stretching them in a frame, removing the hair, scraping and smoothing. Calf, goat and sheepskins were mostly used. The preparations are similar to those for making leather. Thus this writing material — perhaps the oldest known — is structurally quite different from both papyrus and modern paper. Pergament, by its nature, is a finished writing material with a closed and uniform surface.

Today pergament is still produced in small quantities principally for book binders or for special diplomas or certificates. Even during the Middle Ages this high quality material had reached its limit of consumption — it was simply too expensive.

Bark of trees used in Central America

The bark of various trees was used by the early peoples of Central America for writing and painting. The Aztecs and Mayans — major Indian cultures of Central America — made a type of paper from the bark of several species of fig trees.

It is written that the "branches of the paper trees" (with leaves similar to a fig tree) were first soaked in water, then the outer bark was removed and the inner bark was scraped off. This bark was then cut into strips, worked again and built up into a sheet.

Today, paper is still made this way by an Indian tribe in Mexico. It is called Otami paper. It appears that this process was developed at about the same time as the discovery of papyrus.

China, the birthplace of modern paper

The earliest means of recorded communication in ancient China was a form of knot writing. People used knotted cords to count and transmit messages.

In earliest historical times, people used notched sticks for contracts and other purposes. Writing has also been found on bones and turtle shells. During the Chou Dynasty (1122—247 BC), the most commonly used writing materials were bamboo tablets and pieces of wood. Like the clay tablets of Egypt, these were hard to handle.

In the 3rd century BC a new writing material was introduced: pure silk. Brushes and pens were used to paint characters on silk bands, and the Chinese began to write all their books and scrolls on silk.

Using silk created a lot of waste, and so people tried to find ways of making something from it. The waste was soaked and beaten in water into a fine, smooth pulp. This was then poured over a mat made of sack linen stretched on a frame. The fiber settled as the water drained through the mat and thus, after drying, a sheet was formed. Naturally, the structure of this fiber sheet was quite different from that which we call paper today. But it was a beginning of the idea that a writing material could be made from a formed mat of fibers.

Silk fibers, unfortunately, do not have the right properties for good papermaking: they do not swell enough in water, and thus they do not bond together well enough to form a strong sheet.

In any case, neither the bamboo tablets nor the "silk paper" were really suitable and cheap enough to serve as a common writing material.

Tsai Lun and his great discovery

It was from this background that Tsai Lun developed his plans to make paper out of tree bark, bast fibers, rags and even old fish nets. He reported this to the Emperor in the year 105 AD, and he received great honors.

Tsai Lun improved the process of making a thin and uniform sheet: instead of using a rough fabric as the forming mat, he used fine bamboo strips laid alongside each other with hair or silk threads woven crossways to hold them together. This whole mat, or forming fabric as it would be called today, was held in a wooden frame. This "forming frame" was dipped into a tub containing the fibers dispersed in water.

When the forming frame was pulled out of the tub, or vat, retaining strips around the four sides of the frame prevented the paper slurry from running off the frame.

The fibers could thus settle onto the mat as the water drained through. The mat, or forming fabric, was then separated from the freshly formed sheet and was used to form the next sheet.

(Recent archeological findings indicate that paper or paper-like material may have been known somewhat before Tsai Lun's time — as early as 73 BC. However, these findings do not diminish the role of Tsai Lun in the development of papermaking.)

In addition to the development of the forming frame and forming fabric for multiple use, Tsai Lun's idea of finding suitable fibers for making paper sheets was also new. These principles are still applied today.

Tsai Lun was born in Kwei-Yang in southern China; he entered the service of the Emperor Ho in the year 75 AD. In 89 AD he was appointed administrator of the

FIGURES 5 and 6: Papermaking from bamboo is shown in these early Chinese prints. At left the bamboo is cooked in an oven to separate the fibers. At right, sheet forming from a square forming vat.

FIGURE 7: Tsai Lun invented the art of papermaking, according to Chinese records. He taught the preparation of plant fibers and rags by beating with a wooden hammer in a stone mortar. And he introduced the forming frame.

Imperial arsenal. In his free time he worked hard on his discovery of paper, and in the year 114 AD the Empress rewarded his long service with an honorary title and many priveleges.

It is not certain where Tsai Lun made his famous discovery. One report says he lived near Lei-Yang in Honan Province, where a mortar stone was found in which the fiber stock had supposedly been beaten. Another report says that Tsai Lun lived in Tsao-Yang in Hupeh Province. There is a small pond nearby called Tsai's Pond, and it is here that he supposedly first made paper from old fish nets. His art was passed on to the people of this region, according to legend, and even today there are said to be many papermakers around Tsao-Yang.

There is abundant proof of the papermaking of Tsai Lun's time in the form of thousands of pieces of paper from this region near the Great Wall. Tests have shown that the raw materials were the bark of the mulberry tree, Ramie or grass. But rag-based papers have also been found among old Chinese papers.

The art of papermaking first reached Japan in the year 610, having come through Korea. By the end of the 9th century, papermaking was commonly known throughout China and the orient.

Paper's long road westwards

Although papermaking was well known throughout the Middle Kingdom of China, the Chinese managed to keep the secret of papermaking to themselves until the 8th century. This is true despite the fact that pieces of paper had found their way outside the Empire long before then.

In 1901, a rider accompanying Sven Hedin's expedition lost his way near the dry lake of Lop Nor in what is today Chinese Turkestan. There he found the remains of houses in the desert sand: it turned out to be a Chinese military garrison from the 3rd century AD, along the old silk road to Arabia. Many well-preserved pieces of paper were found here bearing Chinese characters. They were

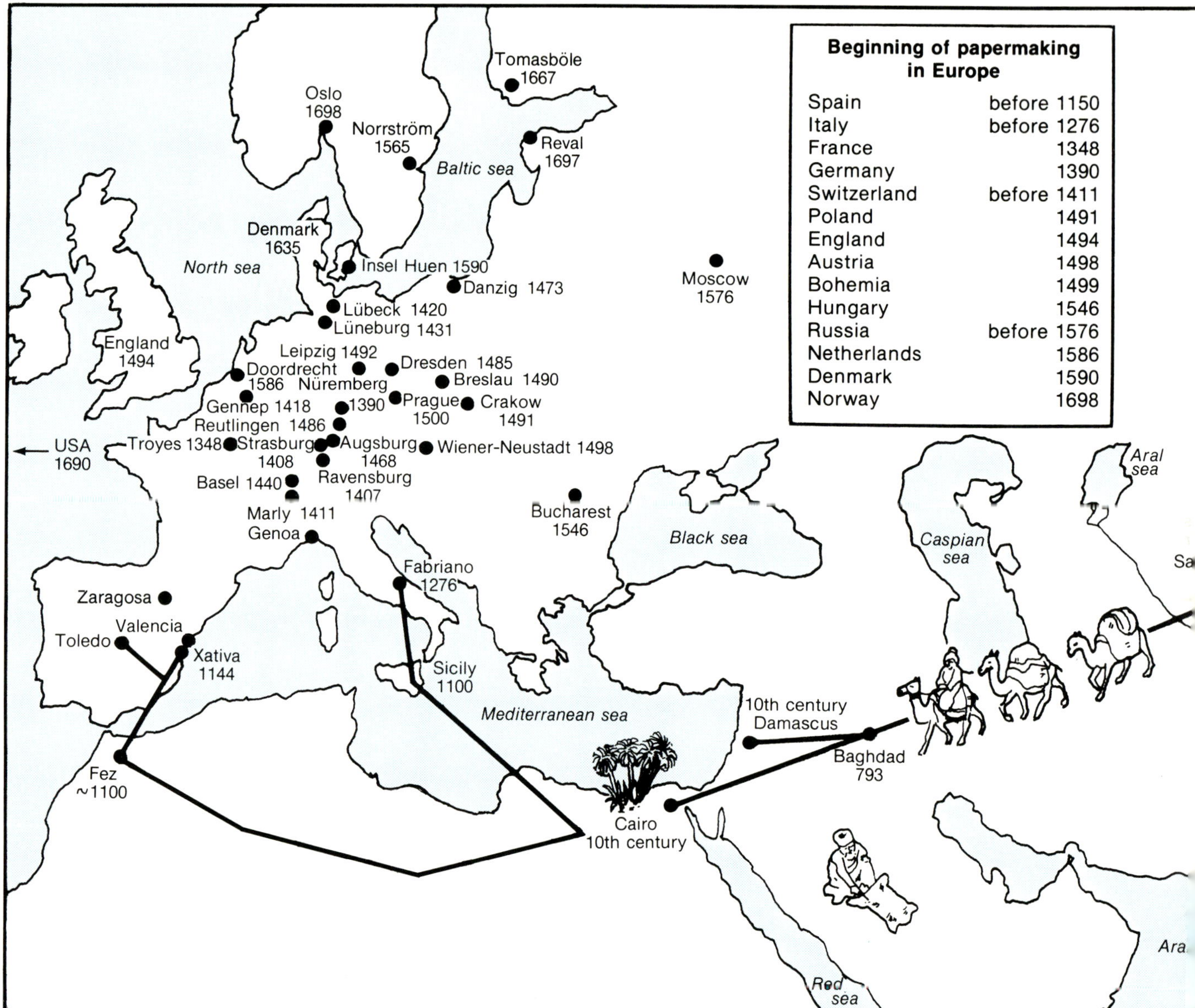

Beginning of papermaking in Europe
Spain before 1150
Italy before 1276
France 1348
Germany 1390
Switzerland before 1411
Poland 1491
England 1494
Austria 1498
Bohemia 1499
Hungary 1546
Russia before 1576
Netherlands 1586
Denmark 1590
Norway 1698
Tomasböle 1667
Oslo 1698
Norrström 1565
Reval 1697
Baltic sea
Denmark 1635
North sea
Insel Huen 1590
Danzig 1473
Moscow 1576
Lübeck 1420
Lüneburg 1431
England 1494
Leipzig 1492
Dresden 1485
Doordrecht 1586
Nüremberg
Breslau 1490
Gennep 1418
1390
Prague 1500
Crakow 1491
Reutlingen 1486
USA 1690
Troyes 1348
Strasburg 1408
Augsburg 1468
Wiener-Neustadt 1498
Ravensburg 1407
Basel 1440
Marly 1411
Genoa
Bucharest 1546
Black sea
Caspian sea
Aral sea
Fabriano 1276
Zaragosa
Valencia
Toledo
Xativa 1144
Sicily 1100
Mediterranean sea
10th century Damascus
Baghdad 793
Fez ~1100
Cairo 10th century
Red sea

Paper's journey from China to the East and West

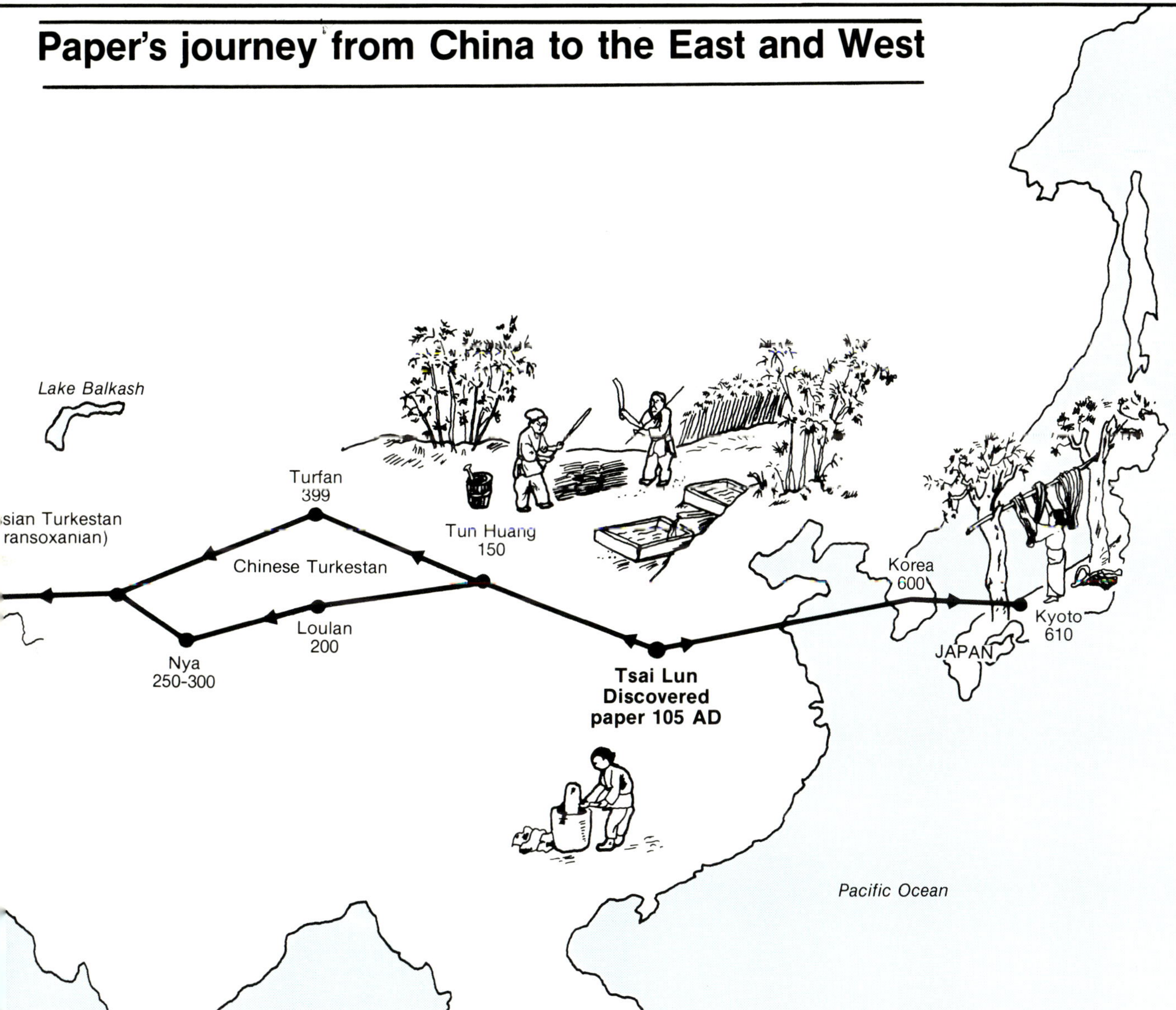

dated from about the year 200 AD.

In the year 751 AD, the Chinese were defeated by the arabs at Transoxanien, in what is today Soviet Turkestan. Among the prisoners the Arabs brought back to Samarkand were several Chinese papermakers. They were allowed to practice their craft, and fortunately Samarkand offered all the necessary raw materials for papermaking — water, linen and hemp were all readily available. And thus papermaking entered the Islamic world.

The Arabs spread the new craft quickly. By the year 794 paper was being used by the Califs of Bahgdad. The Arabs also improved the papermaking process.

(Whether it was the Arabs who first used metal wires to make a forming fabric, or if this discovery — which made the watermark possible — should be credited to the Italians, is not known.)

It is not until the second half of the 10th century that Arab sources speak of papermaking and the use of paper. This is some 250 years after the art had become known. It is also about this time that Egyptian production of papyrus ceased — displaced by Arab paper. During this time, Samarkand remained the center of the Arab world's papermaking.

The main users of paper in those days were the educated, and they often made paper themselves. There were already many grades of paper in use, among them lower quality packaging papers, and the making of colored paper was known.

By the 12th century there were standardized paper formats, and regulations existed covering what type of paper was to be used in documents, invoices and letters.

In the following centuries Islam ruled the Mediterranean, and this proved an effective barrier to the spread of papermaking and paper articles towards the West. During this time, pergament was still being produced in Europe.

It is not fully known what fibers were used by Arab and early southern European papermakers. It is believed that linen and hemp waste were the main fibers, but cotton was also used.

Following the Moors to Spain

Paper now followed the Moorish culture slowly westwards. (See the accompanying map, which traces paper's introduction westwards from China.) The first European paper was made about the year 1144 at Xativa near Valencia in southern Spain. In the year 1260 production and use of paper in Christian Spain began under King Alfons X; paper began to compete with pergament.

Paper was introduced to Italy about 150 years after it had reached the Iberian peninsula. The first paper mill mentioned was Fabriano in the Province of Ancona about the year 1260. Paper, however, had previously been introduced to Italy by Arabs via Sicily. Thus paper entered the Romano-Germanic culture at the end of the 13th century.

Whether papermaking was subsequently introduced to France from Italy or Spain is not well proven, but the first paper production cited in France was at La Pielle near Troyes.

Germany's first paper mill was that of Ulman Stromer, set up outside the gates of Nürnberg in 1390. Fortunately, Stromer wrote down the history of his paper mill: the mill was called "Gleismühle", and it became a major enterprise. Stromer had acquired the necessary skills and know-how through trips to Italy.

The earliest pictorial representation of a papermill that we have is that of the "Gleismühle" found in the "Schedelsche Weltchronik" of 1493.

Papermaking spread quickly through central Europe in the early 15th century. In 1398 there was a concession given for Chemnitz in Germany; in 1407 a mill was built in Ravensburg and later mills were reported in Lübeck and Strasbourg, then in Marly, Switzerland, in 1411. By the middle of the 15th century, the Swiss city of Basel had become an important center, with numerous mills in the nearby village of St. Alban. Many old buildings may still be seen there today.

Use of paper grew slowly

Although the supply of pergament became increasingly difficult and the use of papyrus had ceased entirely in the 12th century, paper's acceptance grew rather slowly. There were, apparently, two main reasons for this: general mistrust of the less durable paper in comparison with pergament, and paper's Arab origins.

FIGURE 9: An early paper mill of 1661.

The fanaticism of Christianity at that time, with its desire to destroy anything that had any connection with Moorish culture, hindered the acceptance of paper.

Paper was, at first, expensive; and this kept its use from spreading quickly. Paper cost less in the East and its cost rose the further west one went. The cost situation improved when more linen began to replace wool for underclothes in Europe — the used linen became a valuable raw material for paper.

Finally, after the 15th century, paper achieved its breakthrough: mainly because it was more flexible and more economical than pergament.

The development of printing, which was greatly advanced at this time by the invention in Europe of movable type and Gutenberg's printing press, was probably the most significant impetus for the development of the paper industry. Printing provided for consumption of previously unimagined quantities of paper.

By the year 1600 there were over 200 paper mills in the German states, of which one-quarter were in Bavaria in the south.

The first English papermaker was John Tate, who started in Stevenage (Hertfordshire) in 1494.

The first known paper mill in Sweden was built at Norrström near Stockholm in 1565. Finland's first papermaking was done in 1667 at Tomasböle in southwest Finland, and the first Norwegian paper mill stated in 1695 in Oslo.

In Holland, the introduction of paper was relatively late. The mill at Dordrecht is first mentioned in 1586. The abolition of the Edict of Nantes in 1685 brought many French papermakers to Holland, which led to an increase in activity.

Paper finally reached America at the end of the 17th century. A German immigrant named Wilhelm Rittinghausen arrived in the colony of Pennsylvania from the Netherlands. He built his paper mill in Germantown, near Philadelphia, in 1690. (Germantown is today part of the city. Rittinghausen's name was anglicized to Rittenhouse and is commemorated in the city.)

The circle was closed. Paper's long road had been

traveled. In some 1,700 years, paper had come from China through the Middle East and Europe and finally America.

At the close of the 18th Century, the Frenchman Louis Robert made the invention which brought papermaking into the industrial age. In 1799 he introduced the first continuous paper machine.

The life of the early papermaker

The early papermakers had to endure a hard life. From dawn until late into the evening they stood at their vats and formed sheet after sheet. Their mills were usually right next to a clear brook or a small river, whose flow was used to power a waterwheel to run the hammermills which beat the fiber. The papermaker had to make his own vats and presses of wood, and above the workshop he built a high tower for drying the sheets. On flat land, where there were no fast flowing streams, many paper mills were driven by windmills.

The rag collecter delivered the necessary raw material, or the papermaker did his own collecting.

The fresh sheets of paper were pressed between cloths and hung on lines for drying. Paper for writing was then dipped in the sizing solution, dried again, and then

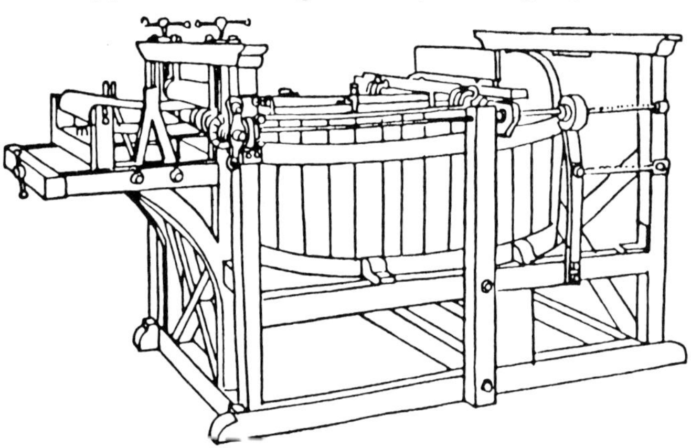

FIGURE 10: The first paper machine, built by Robert in 1799.

calendered or smoothed. It took great effort and dedication to get an old paper mill working smoothly and producing good quality paper. Every papermaker had his secrets, and old traditions were closely followed.

Different papers had distinctive characteristics. The quality of the water played an important role, as did the type of rags used; so did the type of forming frame. Later came the "watermark", the trademark of the papermaker.

Papermakers considered themselves somewhat apart from the other artisians and craftsmen of the times. They called their newly developed craft an art, and instead of joining the usual guilds during the Middle Ages, they joined artists' associations.

Papermaking indeed came more an more to be recognized as an art. This led, in turn, to promoting the construction of new paper mills and favoring them through various privileges. Several universities also had their own paper mills, and since universities and churches needed paper in greater quantities, papermakers came to receive honors and added respect.

The papermaker had to fulfill many requirements: the paper had to have a good "feel" to it, it had to be smooth and white and easy to write on. Printers wanted a somewhat softer quality for better reproduction of the type characters.

The apprentice papermaker had four long, hard years to serve in learning his trade. In those days he was completely dependent on his master. There were no wages — the apprentice received free room and board in his master's house, and the master supplied his clothes. At the end of his apprenticeship, the young papermaker set off to find work in another town, often settling in the home of that town's papermaker.

In addition to the hard work, the papermaker had to live with many risks. Fires destroyed many mills, as did storms and floods, as the early mills were always close to water.

Machine papermaking finally drove the old craft virtually into extinction as mill after mill closed during the 19th Century, bringing to a close the colorful era of hand papermaking.

But even today, more than a century later, there are a number of hand papermaking operations still active in Europe.

FIGURE 11: Watermarks were the trademarks of the early papermakers.

Paper develops through the centuries

Paper may be defined as a flat sheet formed from fibers dispersed in aqueous suspension, pressed and dried with the use of additional heat.

It is the molecular attraction of the fibers which mainly holds the paper sheet together and gives it strength, while the fiber's colloidal properties — basically related to being able to swell in water — which give paper its elasticity.

Many different raw materials have been used to make paper in many countries throughout the years. The earlier papers — according to the reports of Tsai Lun's work — were made of the bark of trees, bast fiber, hemp and old rags. Today, although rags are still used for some very high quality papers, the modern fibers used for papermaking are chemical woodpulps (chemically processed wood fibers) and mechanical pulps (made by mechanically separating the wood fibers).

Bamboo was difficult to pulp, or prepare into usable fiber. The young stalks were cut into pieces and softened by soaking about three months in water. Removal of the green bark exposed a fiberous material that was then cooked 8 or 9 days in a mixture of water and limestone and later beaten with heavy hammers.

A similar, laborious procedure was used for rice straw; it was also pulped by cooking in a limestone solution. Mulberry bark was used mainly in Japan, and was prepared by a highly developed process which separated out the usable fibers. Today, paper made from this fiber is called Kodsu and ranks among the finest Japanese qualities.

The Arabs who learned their papermaking from the Chinese, knew only the use of old rags, as both linen and hemp were available in and around Samarkand. But as demand for paper grew, there was no way that the supply of rags could keep pace; a new way had to be found to overcome the constant shortage of papermaking fiber.

Discovery of woodfiber's potential

The oldest papermakers — far older than mankind — were discovered by the French physicist Réaumur. Several species of wasps and hornets build their nests from a greyish paper-like material. It was proven that these insects in fact used wood and plant fibers to produce their "paper".

FIGURE 12: A papermaker working at his forming vat, from a print dating from 1689.

Many scientists worked on the question of how to make paper without rags. Finally, it was the scholar Christian Schaeffer who recorded his systematic experiments in six volumes entitled: "Trials and samples of making paper without rags or with only small amounts of rags", published in 1765—71 in Regensburg, Germany. Many years passed, however, before Schaeffer's suggestions were put into practical, commercial use.

In 1840, Friedrich Gottlob Keller, an artisan from Hainichen, Germany (in the State of Saxony) conducted trials at making paper from wood. He, too, had been inspired by observing wasp nests, but he got his real idea from watching children make necklaces of cherry pits. The cherry pits were put into a small depression in a wooden board so they could be held and filed down on each side to permit drilling a hole to string the necklace thread. A filing stone was used with water and in the process some of the wooden board was also ground away; in the form of milky white drops it fell from the board and collected on the floor. After drying, there was a paper-like sheet.

This is, thus, the basic process for mechanical reduction of wood with a grinding stone to produce groundwood mechanical pulp which Keller saw. But Keller had no money and had to sell his idea to a machine manufacturer named Völter, and it was 1857 before the first applications of groundwood in papermaking took place.

Groundwood, as this mechanical pulp is still called today, had a grey-white color, was too brittle to be used alone for papermaking. An improvement was needed.

Just as the fibers are mechanically reduced in the groundwood process, in chemical pulping the fibers are extracted chemically from the wood. Wood, just as all other plants, consists of cells. These are surrounded by cell walls which contain various substances not found in the cells themselves. The basic chemical pulping process is essentially to dissolve these cell walls and release —

undamaged — the fibers inside the cells. For reference, an undamaged spruce fiber is about 4 mm long.

Straw was used before wood for papermaking — straw paper was known in France around 1780, though the straw fiber was mechanically reduced and added to the rag-based paper stock. Then, in 1854, the Frenchman Mellier discovered that by cooking the straw under pressure in sulfuric acid, he could dissolve the cell walls to such an extent that the pure fibers were released.

Similar trials were made in England in 1857 by Houghton, and by 1864 the Englishmen Watt and Burgess received a patent in the United States for production of sulfate pulp.

Fiber production processes have been developed tremendously since the early days. Many species of wood and some annual plants are successfully used; chemical and mechanical pulping techniques are highly efficient and fully automated in many cases. Yet the principle of sheet formation from fibers on a woven sieves by dewatering has been retained, and can be seen in the most modern paper mills today.

FIGURES 13 and 14: Hammermills were used to prepare the fibers. At left is the interior of an early hammermill, driven by the waterwheel shown at the right.

Chapter II

THE ART OF HAND PAPERMAKING

This chapter describes the hand papermaking process in complete detail. Everything you need to make good handmade paper is explained here, with a careful description of the actual techniques employed. If the equipment you need is not readily available, drawings at the end of this chapter show how to make your own forming frame and paper press.

What you need for good papermaking

Papermaking needs water, so the best place to do your home papermaking is in the laundry, the kitchen, bathroom or cellar – wherever you have a good supply of water. If it is warm outside, the porch or garden terrace is a good place. A garden hose is a good water supply, but a sink or drain of some sort is very useful.

The work bench or table should have a flat and waterproof surface–preferably a smooth plastic surface. A second table nearby is useful to put things on.

The papermaking tub, or vat, which holds the mixed pulp, should be put on a solid support – strong wooden or plastic crates will do, as will two sturdy stools or saw horses.

Papermaking kits

A specially designed Papermaking Kit may be available at your local hobby store. This kit will contain all the essential ingredients for good home papermaking, including a forming frame, pressing felts, specially prepared pulp, filler clay, sizing and dyes.

Paper presses are normally not found in stores, and the ambitious papermaker should take the time to build his own press. Design and instructions are given in Chapter III. A good paper press is not difficult to build, and it will improve the quality of your paper as well as making the work easier.

Table 1: Equipment and accessories for hand papermaking

Quantity	Description
1	Tub or basin: capacity 60 liters (15 US gallons). This can be of wood, galvanized steel or plastic. An oval form is best.
1	Forming frame: made of wood or plastic, complete with forming wire. (Complete description and instructions for making this are given in Chapter III
3	Plastic buckets: 10-liter (2.5 gallon) capacity.
1	Glass bottle with screw-on top. Capacity 0.5 liters.
1	Small shot glass or small measuring beaker.
2	Pressing plattens or boards: 40 x 40 cm x 0.6 cm minimum thickness (18 x 18 x 8 inches). Hard plastic, wood or hard-surface fiberboard with a waterproof surface. With wood, use a greater thickness.
1	Stirrer or mixer: manual or electric.
10	Absorbent pressing felts. The more the better. These may be filter cloth, cotton or non-woven.
10	Cotton towels (or more): 40 x 40 cm. White dish towels will do.
20	Cardboard sheets (white), 40 x 40 cm, 2 mm thick
1	Clothes line or drying rack.
1	Scale capable of weighing in grams (or ounces).
1	Kilogram (2.2 pounds) of pulp: preferably this should be pulp specially prepared for hobby papermaking, in sheet form.
250	gms White filler clay. Fine powder China clay.
50	Milliliters (cm3) of concentrated paper sizing.
3	Bottles of special dye for coloring the paper.
5	Meters (15–20 feet) of bronze watermark wire, 0.15 mm diameter (0.004 inches).

Note: Some of the equipment or materials listed in the table are not always available, or can only be found in special papermaking kits. However, good home papermaking can be achieved with just the basic materials – dyes and watermark wires are not required.

Basic tools needed: scissors, pliers, screwdriver, pencils and a ruler.

Optional tools and accessories

1	Kitchen mixer.
1	Electric drill with a paint-mixer attachment.
1	Paper press. (Full instruction for building this are given at the end of this chapter on page.
1	Dryer: either an electric clothes iron or a heated platten.
2	Metal sheets – about 25 x 25 cm by 0.4 cm thick (10 x 14 x 3/16th inches) – of aluminum, copper or brass.
2–3	Weights of 5 or 10 kilograms (10 or 20 pounds each.

Preparing the pulp

Many different fibers are well suited for papermaking, and are used industrially today. Woodpulp is the most widely used fiber, and is the easiest and best to use for home papermaking. (This is also known as chemical woodpulp, because it is made by a chemical treatment which separates out the white cellulose fibers from the rest of the wood.)

The woodpulp that is recommended is a specially prepared mixture of fibers from both coniferous (evergreen) and deciduous (hardwood) trees. The pulp is specially refined to produce optimum properties in your paper, such as strength and opacity.

To begin, weigh out about 400 grams (1 pound) of the pulp. Put the sheets of pulp into a bucket with enough water to cover them. It is best to let the pulp soak overnight in the bucket.

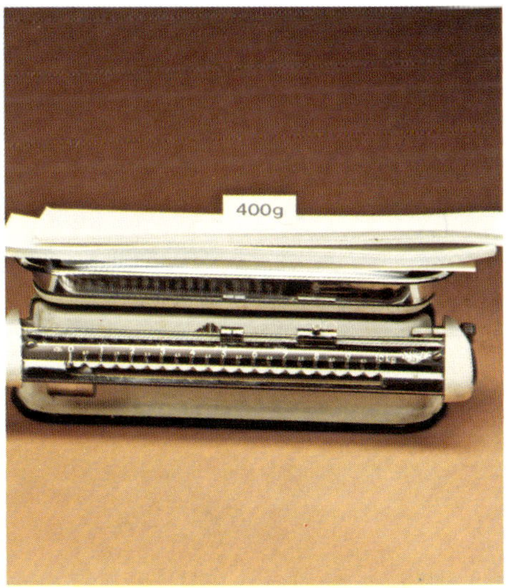

FIGURE 1: Weigh about 400 grams (1 pound) of pulp on a kitchen scale.

FIGURE 2: Tear the pulp sheets into small pieces.

FIGURE 3: Fill the bucket about half-full (5 liters), covering the pulp completely. Leave the pulp to soften.

After soaking, tear the sheets into small pieces, then stir vigorously with a cooking spoon or stirrer until you have a uniform fiber slurry.

How to test the pulp

It is good to check the quality of the pulp slurry – pulp stock, as the professionals call it – before proceeding further. Take a teaspoon full of the stock and put it into a wide-mouthed glass bottle of roughly 0.5 liter size (1 pint). A marmalade jar will do. Fill the bottle about 3/4 full with water, put the cap on tightly and shake briefly but vigorously.

Now hold the glass up to the light. You should see many small fibers floating freely. You may also see small lumps or bundles of fibers – these are a sign that the pulp has not been mixed or "beaten" well enough for good papermaking. These lumps of fiber would show up as spots in the finished paper.

If you see these fiber lumps, pour the contents of the bottle back into the bucket and continue vigorous mixing. Check periodically with the method described

FIGURE 4: Stir vigorously, adding more water to produce a lump-free slurry.

FIGURE 5: Mixing is much faster with an electric drill fitted with a paint-mixing attachment. (Note: protect the drill from moisture and keep your hands dry.) A kitchen mixer can also be used.

FIGURE 6: Use a bottle to test how well the pulp has been mixed or "beaten". Put a teaspoonful of "thick stock" into the bottle, fill it with water, shake vigorously and then see if any lumps or "flakes" are still present.

here to see if the lumps have disappeared.

An electric drill, with a mixing attachment – such as a paint mixing paddle – makes the job of mixing the pulp much easier and faster. A kitchen mixer or blender also does a good job, and in fact is the fastest way to beat the pulp. One minute in the blender is usually enough. But always check the results with the glass bottle. Most blenders have a capacity of about 1 liter (1 quart), so about 8 blender fillings are needed to prepare the full bucket of pulp stock.

Papermaking additives

Your bucket of well-mixed pulp stock has 400 grams of fiber in about 8 liters of water. This can be expressed as:

$$\frac{400 \text{ grams}}{8000 \text{ cm}^3} = 0.05 \times 100 = 5\,\% \text{ consistency}$$

At this 5 % consistency various papermaking additives are now mixed into this "thick stock". Later the stock will be diluted with more water for good papermaking.

The additives used vary with the end product desired, but for best results they all should be blended in the order given here.

Filler clay gives paper a more "closed surface" – better for writing – plus higher whiteness and better smoothness. Add about 10 % clay, based on the dry weight of the fiber: 10 % × 400 grams = **40 grams of clay.**

FIGURE 7: Add about 10 % filler clay (fine white porcelain clay) to the prepared pulp stock. For 400 grams of dry pulp you need about 40 grams of clay. Pre-mix the clay first with water in a half-liter bottle, shaking well.

FIGURE 8: Add the pre-mixed clay slurry to the pulp stock.

FIGURE 9: For colored paper, use a concentrated dye — in this case blue. Intensive coloring is done with only 1 % of dye based on fiber weight. (For 400 grams pulp, only 4 grams, or 4 ml, of dye.) First dilute the dye about 10:1 with water in a small bottle before adding it to the pulp. (Rinse the measuring glass well after using dyes, and wear rubber gloves to prevent stains.)

Weigh out the clay, then mix it first in a glass with fresh water with a 10:1 ratio. The pre-mixed clay slurry is then added to the pulp stock in the bucket.

Dye should be available in concentrated form in the basic colors yellow, red and blue. About 1 % by weight liquid dye (again based on dry fiber weight) is enough to produce a deep-dyed paper. In our case, 400 grams of fiber will require: 1 % × 400 grams pulp = **4 grams of dye.**

First dilute about 1 gram (1cc) of dye with water in a small bottle, then add the mixture to the pulp stock while stirring. (Note: These dyes will stain clothing and skin – use rubber gloves if you have them.)

Sizing helps prevent ink from running into the paper, and is added if the paper is to be written on with ink. No sizing is required if you only intend to use a pencil or ballpoint pen, or if the paper is to be printed.

For good results, you need obout 2 % concentrated sizing based of fiber weight (2 % × 400 grams = **8 grams sizing**).

Pour this out into a small shot glass or, better, a measuring beaker from a chemistry set. As with the filler clay, first dilute the sizing 10:1 in water, then add the diluted sizing to the fiber stock while mixing.

Melier fibers are colored fibers which are intended to stand out from the normal fibers in the paper. These can give your paper a distinctive appearance. Melier fibers are typically used in banknotes and various security papers.

If you want these special effects, take one or more small sheets of Melier fibers, soak them in water in a half-liter bottle and shake until the fibers are fully dispersed. Then add the contents of the bottle to the pre-mixed pulp stock and mix thoroughly.

FIGURE 10: Add the diluted dye to the thick stock. The actual color of the paper will be somewhat duller than the stock.

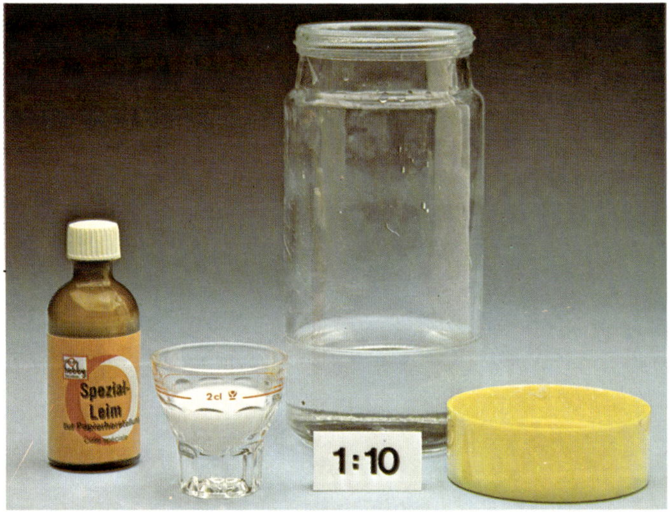

FIGURE 11: Paper must be sized for writing with ink. Use about 2 % sizing based on dry pulp weight, or about 8 grams for 400 grams of pulp.

FIGURE 12: The special sizing compound is first diluted about 10:1 with water and mixed thoroughly.

FIGURE 13: The pre-mixed sizing is then added to the thick pulp stock while stirring thoroughly.

FIGURE 14: Melier fibers (pre-colored fibers) can give your paper a distinctive appearance. Tear up 2—3 small sheets of Melier fibers, put them in a small bottle.

FIGURE 15: Fill the bottle with water and leave the Melier fibers to soak, preferably overnight. Shake well to make sure the fibers are completely dispersed.

FIGURE 16: Melier fibers well dispersed and ready to mix.

FIGURE 17: Add the pre-mixed Meiler fibers to the thick stock. Note: Melier fibers are normally used only with white or very lightly colored papers.

Forming the paper sheet

The forming vat or tub should be set up on its stand, and you should have a table or workbench next to it. Before you start, rinse out the press felts in water, squeeze them dry and lay them one on top of the other on the table.

The forming vat – about 60 liters (15 gallons) capacity is a good size – should be filled about half full with water (30 liters). Take the bucket of prepared thick stock – with all additives carefully mixed in – and add about two-thirds of the contents into the vat or tub. Leave the rest in the bucket.

Thoroughly mix the contents in the tub and add more water until it is about 3/4 full (45 liters). An electric drill with a mixing attachment will do an excellent job of mixing.

Warm water gives better results. It improves fiber distribution in the sheet. In fact, heated forming vats are sometimes built by serious home papermakers.

And now rinse the forming frame, with its top border in place. Also remember, before each sheet is formed you should stir the stock quickly with your hand, because otherwise the fibers will settle.

Study figures 20–23 for good forming technique.

Hold the forming frame with both hands on the short sides and plunge it vertically down into the fiber slurry. At the bottom of the tub, turn the frame horizontal, with the wire facing upwards. The fiber stock should be still and undisturbed at this time.

Raise the forming frame slowly towards the surface. As soon as you see the top border of the frame, use this to align the frame as closely as possible to the perfect horizontal. Then carefully and gently shake the frame from side to side, and backwards and forwards.

Now you can see the fiber layer on the wire, and you should try to get the most uniform distribution

FIGURE 18: The prepared thick stock — with all additives carefully mixed in — must be further diluted for making paper. A plastic tub of about 60-liter (15 gallon) capacity is ideal. Fill the tub about half full with water and add about two-thirds of the prepared thick stock from the bucket.

FIGURE 19: Add more water and mix thoroughly. Use a piece of wood, a cooking spoon, or an electric drill with a mixing attachment. (The pulp stock, even when colored, is easily washed off clothes and skin.)

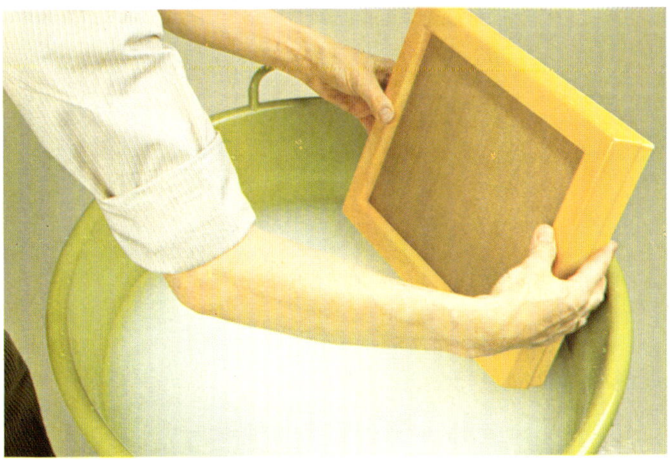

FIGURE 20: First rinse off the forming frame. Now hold the frame — with top border in place — firmly with both hands on the short sides. Lower it vertically, slowly into the tub. At the bottom of the tub, turn the frame horizontal with the wire facing upwards.

FIGURE 21: Raise the forming frame carefully to the surface. When the top of the is visible, shake the frame gently in both horizontal directions — back and forward, side to side — and observe the fiber layer on the wire.

possible. This takes practice, of course.

Raise the frame further out of the stock. Here you will note increased resistance caused by the suction effect under the wire. Let this suction effect work until the water level "drops out" of the forming wire.

Hold the frame a few more seconds to get better dewatering of the fiber mat. Then pull the forming frame completely out, and tilt the frame to let the excess water drip off. Now lay the frame on the table.

Remove the top border carefully, laying it aside without letting any drops of water get on the newly formed sheet. Your first sheet of paper has now been formed.

FIGURE 22: Raise the frame slowly a bit further out of the stock. A uniformly thick layer of fibers should be on the wire. As the frame is raised, increasing resistance is felt. This is due to the suction effect between the wire and the stock in the tub. This helps dewater the sheet further until the water level "drops out" of the forming frame. At this moment, the fiber structure of the sheet is fixed. Allow another 10 seconds for the suction effect to work.

FIGURE 23: Raise the forming frame completely out of the stock. If the sheet is too thin and transparent, remove it from the wire by hand and put it back into the tub, stirring as usual. Rinse the frame off well in running water.

Test sheets – to get good consistency

The sheet of fibers on the wire should now be checked, and any specks of dirt or other impurities removed. The best way to do this is with tweezers.

If the first test sheet is too thin and transparent, no further work should be done on it. In this case, separate the sheet carefully from the wire and put it back into the forming tub, but **be sure to mix it in well** so that no flocks or lumps of fiber turn up in the next sheets you form. Now clean the wire thoroughly under running water.

To get a heavier sheet, take 1 or 2 handfulls of thick stock from the bucket and add this to the tub.

Remember, even if the first test sheet was good, you will always need a reserve supply of thick stock, because each time you form a new sheet, the fiber concentration – or consistency – in the tub decreases, and succeeding sheets would become too thin.

If, on the other hand, your first sheet is too heavy –

if, for example, the fiber layer on the wire is several millimeters thick – then more water should be added to dilute the stock in the forming tub.

With a little experience you will get a good feeling for the right stock consistency necessary to get various sheet weights.

Transferring the sheet

The newly formed sheet of paper must now be transferred from the wire to a felt, a process called "couching" in papermaker's jargon.

Take 2 or 3 of the prepared, moistened felts and lay them smoothly one on top of another on your workbench.

Take the forming frame and place the edge vertically on the end of the felt, then lay the frame – sheet down – on the felt. The sheet should be in the center of the felt.

Now press the forming frame firmly, but evenly onto the felt. Then carefully lift the frame by raising on edge slowly off the felt until the frame is again in a vertical position. Care is essential to avoid bubbles or other damage to the sheet.

Now put the forming frame aside, always with the wire facing upwards.

Lay one or – even better – several of the felts on top of the sheet you have just transferred.

Now go back and form your next sheet – naturally after mixing the stock first quickly with your hand – and couch this sheet onto the next felt.

Soon you will have a stack of sheets couched between felts, ready for pressing.

FIGURE 24: Place the edge of the forming frame on the end of a couch felt. Lay it smoothly over onto the felt. The insert shows a sheet with a thin spot.

FIGURE 25: Lay the forming frame, wire downwards, on the pickup felt and press firmly and evenly, especially on the edges and corners. The paper will now stick to the felt.

FIGURE 26: Lift the edge of the frame carefully off the felt. Remove it smoothly to prevent damaging the sheet.

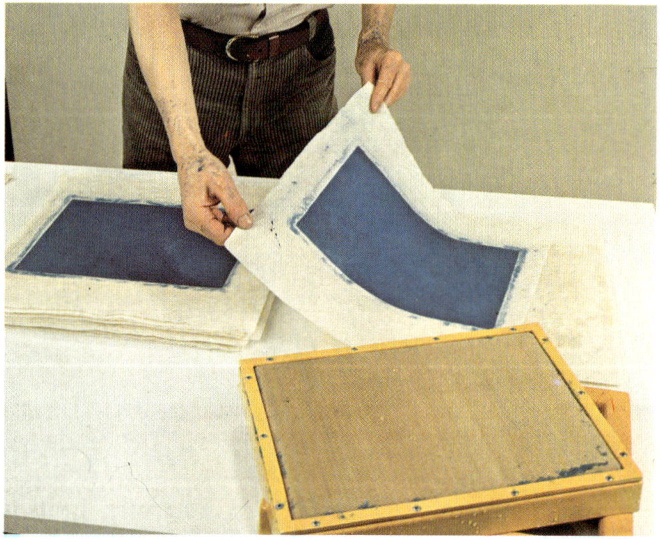

FIGURE 27: After couching each sheet, place it with its couch felt on the table, building up a stack of sheets.

FIGURE 28: When you have formed a series of sheets, put about three felts on top of the stack. Note: it is always good to have a large supply of these felts.

Pressing the paper to remove water

The stack of sheets must now be dewatered as thoroughly as possible to save time and effort in subsequent drying.

The easiest way is to lay the stack on the floor, cover the top sheet with several felts, then put a stiff board on top – or a hard plastic sheet – and stand on it. Water will be squeezed out. (If you do this inside, be sure to place a few old towels around to absorb the water which runs off.)

With this method of pressing, the force is limited by your own weight, and only a few sheets of paper can be pressed at one time.

Dewatering is more effective if you put the sheets between two wooden boards, plastic or even heavy-guage metal plattens and press them together with several big C-clamps.

You can also build your own press. If you plan to

FIGURE 29: The simplest method of pressing is to put the pile of sheets on the floor between two plastic or metal plattens and stand on it.

FIGURE 30: With your own weight, only a few sheets (3 or 4) can be pressed at one time. You can put towels on the floor to soak up the water.

FIGURE 31: The ambitious home papermaker can easily build his or her own paper press. Complete construction details are given in Chapter III.

Using an ordinary hydraulic auto jack, up to 10 sheets can be pressed at one time. Normally, the pile of sheets is pressed twice, the second time with the pressure on for at least 10 minutes. The water that runs out the edges of the pile and onto the lower platten should be wiped off by hand.

continue papermaking seriously, then you definitively should build your own hydraulic paper press. (Detailed instructions are given in Chapter III).

In a good hydraulic press you can put up to 10 sheets – always with their felts, of course. On the press bed, lay a heavy flat wooden board or plastic platten sized slightly larger than the couch felts. Then put on the pile of sheets and put another board or platten on top. Add as many more boards or metal plates as required to make up the necessary height in the press, and to protect the press platten.

Place the hydraulic jack in the middle, line it up with the crossbar of the press and pump it up. Do not forget to allow for the water run-off! When the jack is activated, the pile is compressed. Remember, an ordinary hydraulic jack – found in any auto supply store – can exert forces of 1 to 2 tons. This gives excellent dewatering, much better than was formerly possible in hand papermaking.

After one pressing cycle, it is recommended that you release the pressure, add more wooden boards to compensate for the reduced height of the stack and pump up the jack again. Now leave the jack under pressure at least 10 minutes. Wipe away the water as best as you can from the edges of the sheets. Do the same with any water on the lower press plate, so that this water is not re-absorbed into the sheets when the pressure is released.

After pressing, lay the pile on the table and begin, carefully, to remove the moist sheets. This requires care and skill, but you will get the feel of it quickly. Lift a corner of the sheet with your finger, then get all your fingers under one side of the sheet and lift gently.

Lay the moist sheets on top of each other, preferably on a press felt or a plastic platten, and then put the new pile back in the press – suitably covered with another platten – and dewater further under medium pressure. This gives a better surface and denser sheet formation.

FIGURE 32: Place the dewatered pile back on the work bench or table.

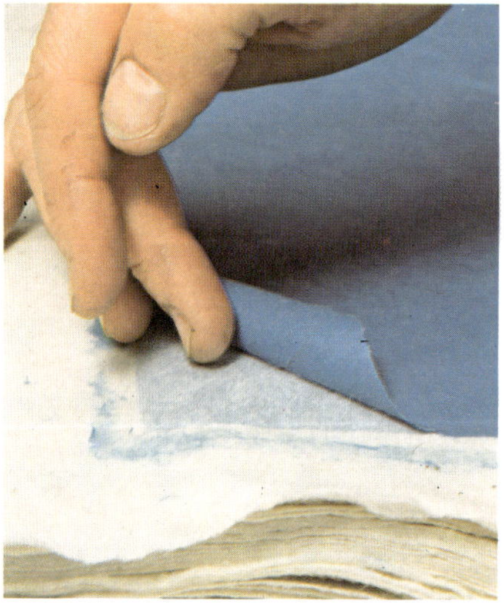

FIGURE 33: Lift up one corner of the sheet, then take the sheet carefully in both hands and lift it off the felt.

FIGURE 34: Lay the moist sheets — without anything in between them — on top of each other for another pressing.

Since the moist sheet is not very strong it should not be carried or moved very far before beginning the next step in the papermaking process – drying.

Drying the paper

The oldest method of drying paper was to leave the wet sheet on the forming wire itself and let it dry in the air. This was hardly practical, for it took very long; there was no pressing (rapid mechanical dewatering) and one could only do as many sheets at one time as one had forming frames. Air drying on plaster walls, on boards or on flat tables or floors were also methods used at one time.

During the last days of hand papermaking in the 19th century, 4 or 5 sheets were taken after processing and hung together on a line. This method is still used today, using a heavy plastic clothesline – if possible with two lines strung close to each other, about 8 cm (3 inches) apart, with the sheets laid over both lines to avoid creasing and stretching of the sheets. In early days, however, the lines were made of horsehair and were coated with beeswax.

In the old paper mills there were air slits in the outside walls to provide a steady flow of air across the drying floor.

Today, much faster drying is possible using heat – for example, with a normal adjustable clothes iron. Lay a cotton towel on the ironing board, then put the moist sheet of paper on it, then another 1 or 2 cotton towels. Put on several cotton towels about 40×50 cm (16×20 inches), as they absorb the moisture well. Adjust the iron to medium temperature and move it constantly on the towels. Check now and then to see how dry the sheet is.

Fast and more uniform drying can be achieved with an adjustable electric heating plate, like those used to warm dishes. If the heating plate is too short or narrow for the paper format, then you can lay a 3-mm thick aluminum sheet on top of it. The ideal size for drying a standard A4 size sheet is about 25×35 cm.

FIGURE 35: The oldest drying method can still be used. Hang the moist sheets on a wash line for air drying. Afterwards, however, the creases have to be pressed and smoothed out.

FIGURE 36: With a clothes iron (preferably with adjustable temperature) you can dry a paper sheet placed between two cotton towels. Keep the iron moving around on the sheet.

First lay a cotton towel on the hot plate. Then carefully put the first sheet of paper on it, covering it with sevaral more towels. On top of the towels place a stainless steel plate, also about 3 mm thick, and weigh it down with weights or bricks. Ideally, the top steel plate should have holes in it to allow steam to escape more easily.

A good drying method is the following: take the dewatered sheets from the press and separate them from their felts (as described previously). Lay them immediately on dry felts, put them back in the press and leave them under pressure for several hours, or overnight.

The still slightly moist sheets-after removal from the press-can then be placed between sheets of clean white cardboard and pressed again for a few hours. They should now be almost dry.

To get a very smooth surface, you can put the paper sheets between plastic film or sheets (such as plastic file folders), and again press them for several hours.

FIGURE 37: Check frequently to see if the sheet is dried enough. Set the iron for medium temperature.

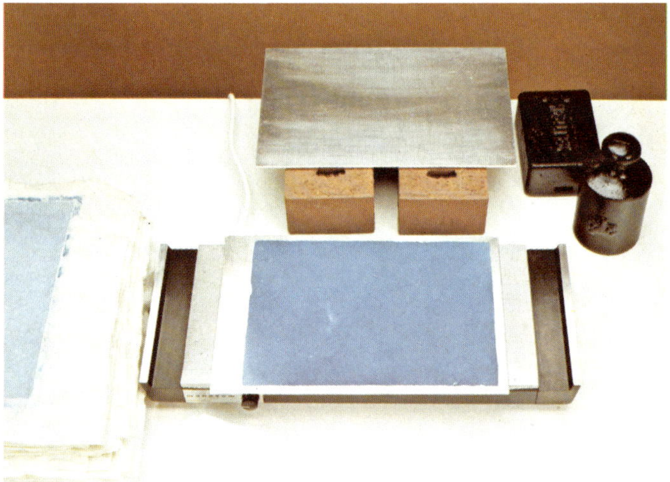

FIGURE 38: Drying is easier and faster with an adjustable electirc warming plate. A sheet of good heat—conducting metal such as copper or brass can be put on the hotplate to get the right size. Put a cotton towel under the paper.

FIGURE 39: To absorb the moisture, put several cotton towels on top of the paper, then put on another metal plate and weigh it down. With medium temperature, a sheet can be dried in about 4 minutes.

FIGURE 40: Check the drying as usual. Be careful of the hotplate.

FIGURE 41: The finished product — your own paper.

Sized paper and the ink test

As mentioned previously, unsized paper is quite suitable for writing with a pencil, ballpoint pen or typewriter, and can also be printed. However, to make paper for writing with ink it must be sized. The sizing prevents the ink from running. The natural absorption of the paper is reduced through sizing by closing the capillary voids between the fibers.

There are several methods of sizing paper. The earliest process was rather laborious: the dried paper sheets were drawn through a sizing solution and then re-dried. One can achieve, today, roughly the same effect with a gelatine solution.

Today virtually all writing paper is treated with rosin size, using the technique of internal sizing developed in the last century by Moritz Friedrich Illig, a watchmaker from Darmstadt, Germany. As the term suggests, internal sizing takes place in the pulp stock before the paper sheet is formed.

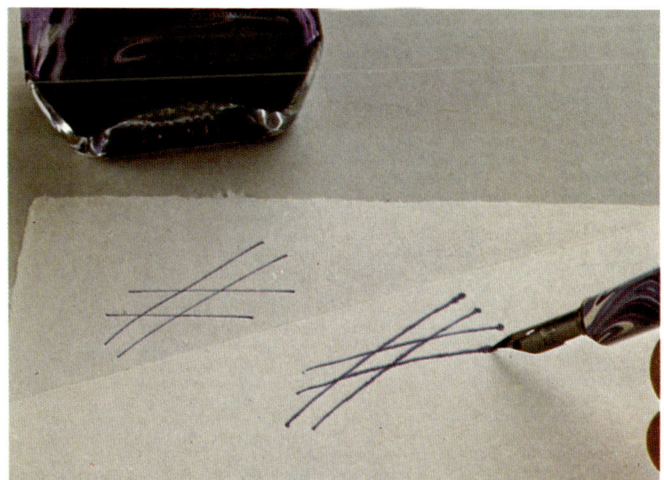

FIGURE 42: The ink test is a check of how well the paper is sized. A well sized paper holds sharp, clean lines. Ink will run on an unsized sheet.

Most recently, synthetic sizing has been introduced in place of rosin size. Synthetic sizing is by far the best way of sizing handmade paper. (See page 24 for description of adding sizing compound to the pulp.)

To check whether the paper has been correctly sized, you can use the simple ink test. With a plain ink pen make a few crossed lines on the paper and see whether the ink "stays" or "runs", particularly at the points where the lines cross. On a well sized paper, the ink will not run (see Figure 42).

Basis weight and fiber concentration

The basis weight of paper is normally expressed in grams per square meter (g/m^2). Some typical basis weights of commonly used papers are give in the table below.

Table 2: Common grades of paper and their basis weights

Type of paper	Basis weight g/m2	Weight of an A4 sheet in grams
Airmail letter paper	15	0.94
Newsprint	45–48	2.8–3.0
Writing/letter paper	80	5.0
Handmade letter paper	100	6.25
Handmade greeting cards	150	9.4

Note: A DIN A4 format sheet is 1/16th of a square meter. So if you weigh a sheet of handmade paper on a letter scale (or any accurate scale), just multiply by 16 to get the basis weight in g/m^2.

The basis weight of a handmade sheet is largely dependent on the concentration of fiber, or consistency, in the forming tub or vat. To predict the basis weight of the sheet you are going to form, you can do a simple calculation. In our example, we assume:

- The forming tub contains 50 liters of water = **50000 cm^3**
- In it are about 2/3 of the 400 grams of pulp you first mixed in, or 2/3 × 400 = **266 grams of fiber**

$$\frac{266 \text{ grams}}{50000 \text{ cm}^3} \times 100 = \textit{0.53\% consistency} = \textit{5.3 grams/liter}$$

- If the border of your forming frame is 15 mm high, then the volume enclosed between the forming wire and the upper edge of the border – assuming standard A4 size frame – is: 21×29.7×1.5 cm = **935 cm^3**
- Multiplying this by the stock concetration of 5.3 grams per liter gives: 0.935 1×5.3 g/1 = **4.95 grams** of fiber

If one considers the additional build-up of fiber due to the raising of the forming frame out of the tub,

FIGURE 43: Samples of colored paper. Green, for example, is made by adding some yellow dye to blue-dyed stock.

then you will end up with about 5 grams of fiber on the wire. This gives a basis weight of about 80 grams, typical for a good writing paper.

To make a 100 g/m^2 sheet, you will need a consistency of about 0.6 % in the tub. This is equivalent to 6 grams/liter or 300 grams of pulp in the 50 liter tub.

Remember, you have reserve stock in the plastic bucket to compensate for the fiber used during your paper production. You know now that each sheet you form removes 5 or 6 grams of fiber from the tub. If you are going to make much more than 10 sheets of paper, or perhaps envelopes as well, then you will want to prepare a bit more than the 400 grams recommended at the beginning of this chapter.

The same amount of pulp is also enough for greeting cards; although they are heavier in basis weight, their format is smaller.

In general, one can say that the orginal "thick stock" must be diluted about 10 times – from 5 % to 0.5 % consistency – for good forming. This is roughly the same consistency used on modern paper machines today.

Fiber recovery and conservation

Normally you will form as many sheets as the prepared fiber stock allows – until it is all used up, including the reserve. This point is reached when the sheets begin to be too light and the insufficient stock consistency in the forming vat (fiber concentration) can no longer be compensated with more reserve stock.

The remaining stock can, of course, be easily washed down the drain with water. However, the thoughtful hobby papermaker will not throw away this valuable raw material. It can easily be recovered and saved for the next papermaking session.

Place an ordinary plastic sieve (a salad drainer, for example) over a bucket or sink. Lay a clean cloth in the sieve. Now pour the remaining contents of the forming vat into the sieve, using another bucket or container if necessary. The water that drains through the sieve can be thrown away.

The fibers collected on the cloth are further dewatered by twisting and squeezing the cloth firmly. The crumpled, moist lumps of fiber can now be stored in a plastic bag.

If you are not going to make paper again in a few days, add a few drops of formaldehyde to the fiber to prolong its storage life. In the case of blue-dyed stock, this can be reused as Melier fibers.

Storage is best in a cool place – the refrigerator is ideal. This recovered fiber may be stored several months this way, and can easily be mixed with new pulp the next time you make paper.

Good clean-up habits

After making paper, thoroughly wash out all the buckets and other containers which have contacted the fiber stock. Cleanliness is important, because any left-over dried fibers can come out as flocks or specks in the sheet the next time you make paper.

The forming frame should be cleaned very thoroughly, but carefully, and packed well for storing in a safe place. It is best to store all equipment, including additives, in boxes so they do not collect dust.

The couch felts should be washed out and dried, particularly if you have made colored paper. Felts which have been in contact wiht colored paper should be kept separate – the color, however faint, may tend to run on the next batch of white paper you make.

If you have used a press, be sure to grease the spindle to prevent rust.

Making envelopes

To learn how to make envelopes, carefully take apart a standard envelope and see how it is folded. (In these examples, we are using a standard format DIN C6 evelope – 162×114 mm.)

Flatten the envelope out on a table (Figure 44); note that it is a square sheet of paper with one corner cut off and four small cut-outs in the sides. The side of the square is 210 mm, equal to the width of the standard DIN A4 letter sheet you have formed.

An envelope can be made easily from a handmade sheet of A4 sized paper by simply cutting to size. First make a cardboard template or pattern with the exact measurements of the envelope. Lay it over the sheet and cut carefully with a knife.

Your envelope (Figure 45) has three handmade paper edges and one cut edge.

If you want your envelopes to show the true handmade paper edge on all four sides, then put a special insert – of rubber, plastic or wood – in the forming frame (Figure 47) so that the sheet formed is square.

As shown here, the insert also includes the one cut corner, so that only the four small cutouts need to be made after forming.

Pressing and drying are done exactly as with full-sized sheets.

If you decide to cut your envelopes from letter-sized sheets, you will naturally want to use sheets which might have some small defect in the area which will be removed.

When gluing the envelope, put a card inside it to

FIGURE 44: A standard envelope. Carefully cut open an envelope and spread it out. Note the cut-outs.

FIGURE 45: Make a master template or pattern out of cardboard to cut envelopes from full-sized sheets.

FIGURE 46: Check the accuracy of your patterns by putting both of them on the forming frame. If you want to make both your letters and envelopes with watermarks, the watermark should be placed in the middle of the envelope area.

FIGURE 47: To form envelopes — with the real handmade edge on all four sides — put an insert into the forming frame. It should be about 15 mm thick, as high as the retaining edge of the frame. Foam, rubber, plastic or wood may be used.

FIGURE 48: Using the cardboard pattern, draw the envelope cutouts on the sheet, then carefully cut them out with scissors.

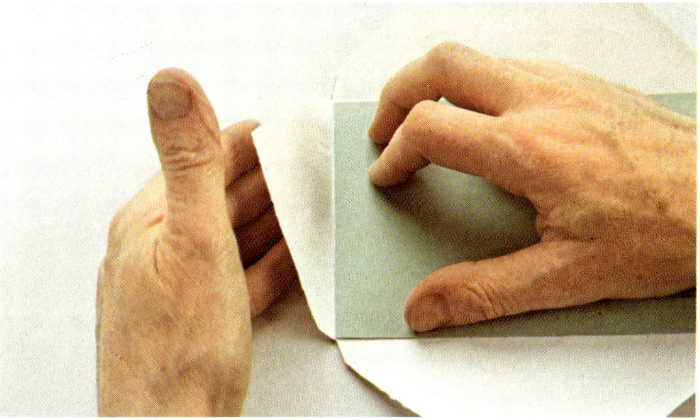

FIGURE 49: Cut out a piece of cardboard the size of an envelope and use it to fold up the sides and ends of your envelope neatly. Now the envelope can be glued with paper glue — put a card inside until the glue dries to keep the envelope from sticking together.

prevent it from sticking together. Leave the card in until the glue dries.

If you want to make elegant lined envelopes, cut out a square of smooth lightweight paper (purchased) about 10 mm smaller on each side than the envelope square and glue in place.

If you intend to make both letter paper and envelopes with watermarks (which are described in the next chapter), then you should place the watermark in about the center of the area of the envelope sheet. To locate the position of the watermark, lay both the cardboard patterns – for the envelope and for the cut-out – in the forming frame (Figure 46).

FIGURE 50: An example of a simple transparent watermark. Formation is good all around the watermark.

Watermarks and how to make them

The simple transparent watermark always appears as a light mark in the paper when the sheet is held up against a light source. This is because the paper is, in fact, thinner at the watermark, a result of less fiber stock in that particular spot. This is achieved by sewing wire figures onto the forming wire.

The top forming wire must be removed from the frame in order to sew the watermark on.

First draw the design of the watermark full scale on a piece of paper. Remember, a fairly simple form will be much easier to execute than a complicated one. Sharp bends are to be avoided in watermark figures, and keep adjacent lines of the figure a bit apart from each other to ensure that enough fibers will form around the wires and not leave a hole in the sheet.

Also avoid crossed lines. A crossed line is always made with two right-angled pieces laid next to each other (see Figure 50). The same principle is used for the letter "T" or the numeral "4", for example. Avoid very small pieces because they are hard to sew onto the forming wire.

Copper or bronze of 0.5 to 1.0 mm diameter is recommended for making watermarks – it is easy to bend and gives a good, clear watermark. Finer bronze or stainless steel wire of 0.15 mm diameter is used to sew the watermark onto the forming wire.

After bending the wire pieces to form your design, remove any burrs or rough edges with a fine file, then lay the pieces down on your drawing to check their accuracy.

Now lay the pieces of the watermark on to the forming wire and holding them temporarily in place with a bit of adhesive tape. The pieces should lie flat on the forming wire. No sewing needle is used – the fine wire is passed back and forth through the forming wire, and always cut and knotted on the underside. Take care that the forming wire is not dented or bent, as this would impair sheet forming.

FIGURE 51: Some typical simple figures and shapes for watermarks, which you can make easily.

When the entire watermark has been carefully sewn on, the forming wire can be remounted in the frame. If the forming wire has any dents or bumps, these can be tied down with fine wire onto the support ribs under the wire.

Normally a watermark is left on the wire. If you want to make a new watermark, it is best to make a new forming frame as well and start with a new forming wire.

Your first watermarks should be simple. Later more complicated watermarks can be made, such as family coats of arms or insignia.

Shadow watermarks

The shadow watermark is more difficult and challenging type of watermark to make. It has darker areas in addition to the light, transparent lines of the normal watermark.

The technique for these watermarks involves embossing the watermark form directly onto the forming wire (See figures 69 and 70). The raised lines are pressed in from the underside of the wire, the indented lines are pressed in from the top side.

During forming, the raised section displace fiber – just as the wire watermark does – and this results in a thinner spot in the sheet. In contrast, the indented lines collect more fiber and produces a thicker and darker area, or shadow, in the sheet.

The lines should be embossed about 3 mm deep.

It is best to practice shadow watermarks on an extra piece of forming wire. Try simple figures, using the back of a knife or a screwdriver as the embossing tool. Try embossing one side first, then both sides.

A rubber or silicone rubber block is the best support for embossing. Otherwise lay the wire on several smooth towels.

FIGURE 52: To make watermarks you need lined drawing paper and a drawing pencil, bronze or copper watermark wire (0.5—1.0 mm in diameter), fine stainless steel sewing wire (0.15 mm diameter), a measuring caliper, pliers, a fine file, scissors and adhesive tape.

FIGURE 53: Dismount the forming frame and remove the forming wire. Take care not to nick or dent the wire.

FIGURE 54: Draw the watermark design in the desired place on the paper format. For envelopes, remember to put the watermark in the right place.

FIGURE 55: Lay the forming wire on the drawing. Since the edge of the wire is flanged, put a 5-mm-high stack of paper under the drawing to support the wire and prevent bending or otherwise damaging the wire.

FIGURE 56: With a felt pen, copy the design on the wire.

FIGURE 57: Check the diameter of the wire with calipers. The watermark wire should be 0.5—1.0 mm in diameter; the sewing wire about 0.15 mm diameter.

FIGURE 58: Following the original drawing, bend and cut the wire pieces as required. Always bend first, then cut when the shape is right. Do not cut too short.

FIGURE 59: File down the ends of the wire to remove burrs. A fine key file or nail file will do well.

FIGURE 60: Lay all finished pieces on the drawing to check them for shape and flatness.

FIGURE 61: Use adhesive tape to hold the pieces temporarily on the forming wire. Leave the ends free for sewing.

FIGURE 62: Cut pieces of sewing wire about 15 cm long. Bend them in the middle and thread both ends through the forming wire around the piece of watermark wire.

FIGURE 63: The size of the loop should not be any larger than the diameter of the watermark wire, so the watermark will be held tightly.

FIGURE 64: From the underside of the forming wire pull the sewing wire ends tight and twist the ends several times. When both ends of the watermark piece are tight, remove the adhesive tape and add more stitches — usually at 10 to 15 mm intervals.

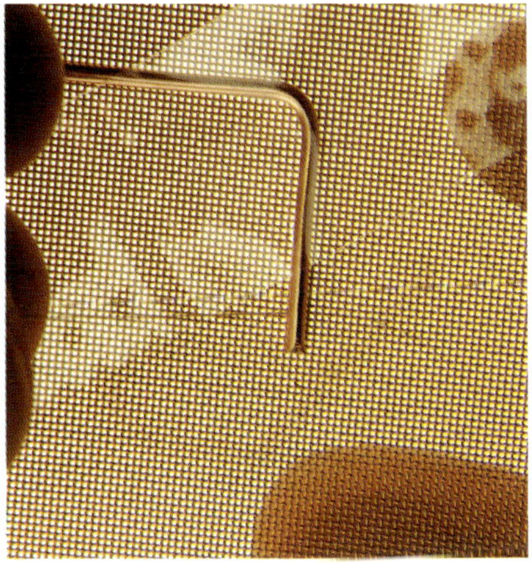

FIGURE 65: After twisting tightly, cut the ends of the fine wire, leaving about 5—7 mm free. This twisted end is then bent to one side.

FIGURE 66: Here is the finished watermark, held firmly but neatly by several stiches of fine wire.

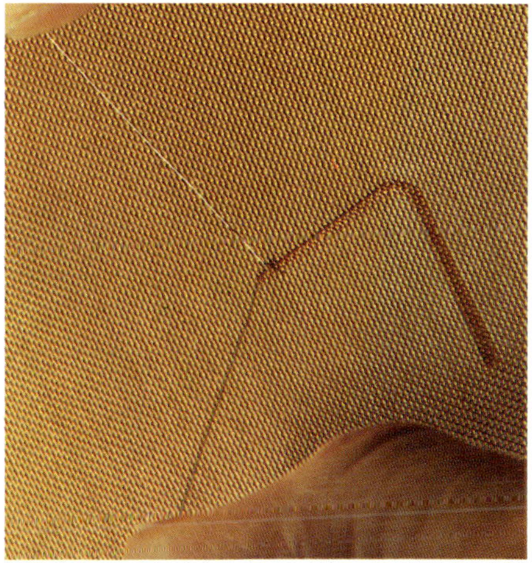

FIGURE 67: Now remount the forming wire in the frame. Make sure all pieces are put back together correctly.

FIGURE 68: During sheet forming, the watermark will be clearly visible — you can see the fibers displaced by the watermark figures.

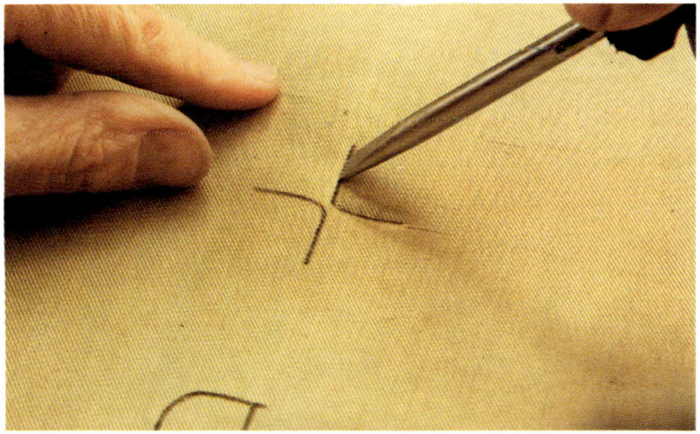

FIGURE 69: To make shadow watermarks with raised sections, the wire must be embossed from the bottom side. The back of a knife or a screwdriver can be used as an embossing tool. To get the correct image, the reverse or mirror image of the watermark should be drawn on the underside of the wire. This can be done against a window. Indented lines are embossed from the top side of the wire. These will appear dark in the finished sheet.

Portrait watermarks

The advanced hobby papermaker who wants an even more ambitious challenge can try portrait watermarks. These should, however, only be attempted if you already have learned drawing, painting or modelling fairly well.

First, you should have a good photograph of the subject to work with. The face or head of your subject in the photo should be about 10 cm (4 inches) across.

The portrait is "engraved" into a wax sheet about 2 cm thick by 20 cm square. In the "engraving" process, the lighter areas are left untouched and the darker areas are engraved – the darker the area, the deeper the engraving. The deepest points should be about 6 mm deep. (The maximum depth may be compared to that of the shadow watermark, where the embossing depth is about 3 mm on each side, or a total of 6 mm.)

In order for both the light and dark areas to stand out from the background of the sheet, the rest of the wax area around the portrait must be removed to a depth of about 3 mm. Now, seen from the median level, we have raised and deepened areas, each no more than 3 mm from the median position. This is equal to a total engraved depth of 6 mm.

The finished engraving is now cast in silicone rubber to obtain a durable mold of the wax form. This silicone mold is, however, still too soft to be used for embossing the forming wire. So, it is put into a casting box and cast in hard epoxy resin – at least 2 cm thick. This resin block becomes the master embossing form.

In addition, a matching opposite form must be cast in epoxy resin. To do this, take the original silicone mold and put this in the casting box and cast it again, this time in silicone rubber. This produces a matching, opposite mold which is then used to cast the matching epoxy form. This in turn becomes the matching embossing die.

Both epoxy dies are now put together and finished so that their bases are parallel.

Embossing of the forming wire is done in a press – a paper press will do very well. The upper and lower epoxy dies are put into the press with the forming wire between them and the set is then pressed together.

If you want to preserve the life of the embossing dies, they can also be cast in metal.

Finally, after carefully checking the embossing and comparing it to the original, the embossed forming wire is remounted in the frame. Small defects can be corrected by hand embossing.

Making portrait watermarks is demanding work, but the results can be beautiful and very satisfying for the talented papermaker.

FIGURE 70: Close-up of the forming wire with a raised line embossed from the underside. This type of raised line will appear lighter in the final sheet.

Chapter III

BUILDING A FORMING FRAME AND A PAPER PRESS

The traditional forming frame for hand papermaking is made of wood. Recently developed plastic forming frames, similar to the wooden frames, permit inexpensive mass production. They also have the advantage that they are absolutely water-resistant and do not warp or swell in water.

This chapter describes construction of a traditional wooden forming frame, as this is the type most home papermakers employ.

The base of the rectangular forming frame is best made of larch, as this wood will not warp when wet. Naturally, pine or exotic hardwoods (mahogany, teak) can also be used. It is best, in any case, to choose a wood which is easily worked, and finish the frame with a coat of good oil.

The outer part of the frame can be glued with a good waterproof wood glue. There is no need anymore to use dowels. Higher strength can be obtained, however, by fitting wooden dowels in the corners.

The ribs built into the frame hold the forming wire flat and support it during sheet forming. When couching the wet sheet onto a felt, these ribs transmit the pressure uniformly onto the wire and the sheet, and prevent damage to the wire. The space between the ribs is where the suction develops to aid dewatering of the fiber mat during forming.

The ribs are always parallel to the short side of the frame. Normally they are mortised into the long sides of the frame and glued in place.

Best spacing of the ribs – centerline to centerline – is 30–35 mm for an A4 sized frame. The top edge of each rib – which actually supports the wire – should be 1–2 mm wide (3 mm is absolute maximum). The ribs are tapered down to a base width of about 10 mm. The small top edge is necessary to assure undisturbed and defect-free dewatering of the sheet.

Using appropriate extruded aluminium profiles for the ribs can save time. An "L" profile of 40×10×2 mm is very good. (See drawing in Figure 4.) These ribs can be quickly sawn to length, pressed into grooves cut into the frame, then fastened from below with brass or aluminum screws to keep them from falling out. The ribs should fit tightly in the frame grooves.

The inside dimensions of the base frame should correspond to the size of the sheet to be formed – in our case the DIN A4 or 210×297 mm. You can add about 2 mm to each dimension for a safety margin; in any case, the top frame determines the actual sheet dimensions.

The top frame or border is normally made from the same wood as the base frame.

You can compare this to a picture frame. There are plenty of pre-cut frame profiles available in art supply stores, and it is recommended that you make the top frame profile by glueing together several pre-cut profiles.

In principle the inner edge of the top frame should sit cleanly on the wire, thus the top and base frames should have parallel sides. There should be enough play between the top and base frames for the top frame to be lifted off easily after sheet forming. Normally about 1 mm in each direction is enough.

The top frame should be well oiled.

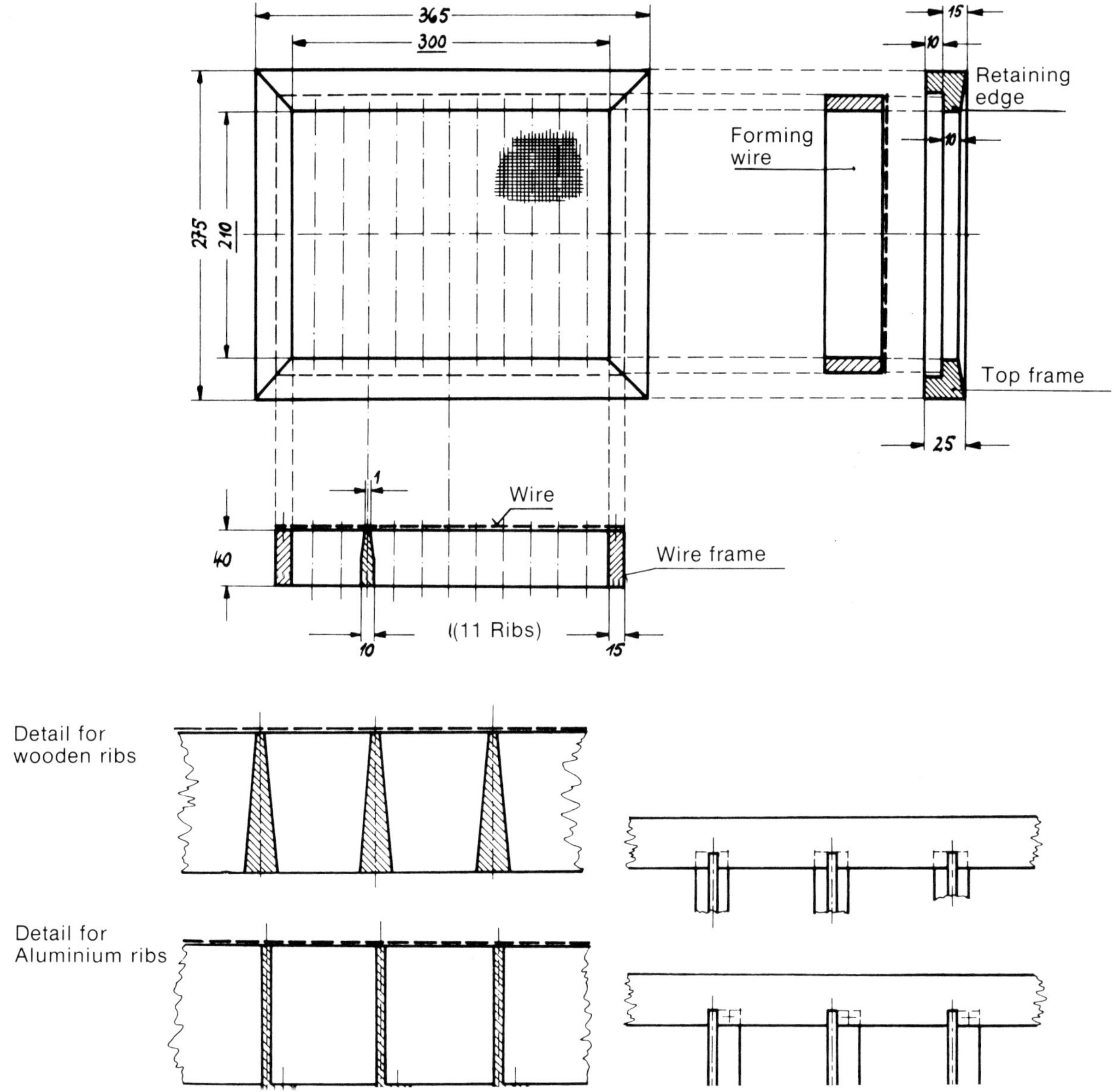

FIGURE 1: The forming frame can be built to these drawings. For larger sheets, dimensions must be correspondingly larger.

Fitting the forming wire

The fine-mesh, non-corroding metal wire sieve may now be fitted to the forming frame. The wire can be of bronze, aluminium, or copper; and should have about 20 wires/cm.

To keep this fine wire mesh flat on the frame and to prevent it from getting dented or bent during use, it must have a supporting mesh underneath it. This support wire, as it is called, shoud be of the same material as the top wire, but with a mesh opening of about 3×3 mm.

The top wire and the support wire are then laid together on the frame and fastened. A 1-cm wide brass strip is laid aroud the outside edge of the wire, and holes are pierced at about every 3 cm. Both wires are thus fastened with brass tacks driven through the brass strip and through the two wires into the forming frame. Fine brass screws may also be used, and the brass strip can be counterpunched to take the heads of the screws.

When mounting the wire, care must be taken that both wires are perfectly flat, otherwise there can be problems in forming and couching the sheets.

FIGURE 2: A forming frame made of larchwood seen from below. The 10-mm wide supporting ribs, which taper down to 2 mm width, are fitted into the long side of the frame. The support ribs are spaced about 30 cm apart.

FIGURE 3: The wooden top frame is mortised and glued.

FIGURE 4: Aluminium angle profiles make good support ribs. Ideal is 40×10×2 mm, shown here. Slots are cut (2 mm wide and 5 mm deep) in the wooden frame at 30 mm spacings. The aluminium profile must fit tightly; it is best to do a test piece first. When the fit of the aluminium ribs has been tested, the frame can be glued. Then cut the aluminium profiles to length, press into place and fasten underneath with brass screws. The top edge of the ribs must be even with the top edge of the frame.

The paper press and its functions

A good press has several functions in papermaking. First of all, a press assures the best possible dewatering of the sheet. At the same time, good compaction of the sheet is an advantage, as the sheet becomes more "closed".

A press permits higher productivity in home papermaking. An entire stack of sheets can be pressed at the same time. And, the more water that is pressed out, the shorter the drying time required.

The original presses that we know of were spindle presses, such as those used to press juice or the type used on early printing presses. At first wooden spindles were used, later steel. Today, for craft papermaking, most people use hydraulic presses, which are far more efficient and permit much higher pressure with far less effort.

The press consists of upper and lower wooden cross-frame members (see Figures 5 and 6). These cross frames are separated by tension rods. (These can be found in many hardware stores.)

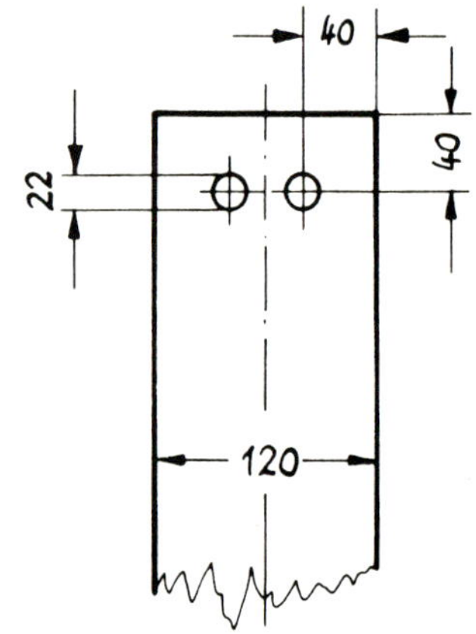

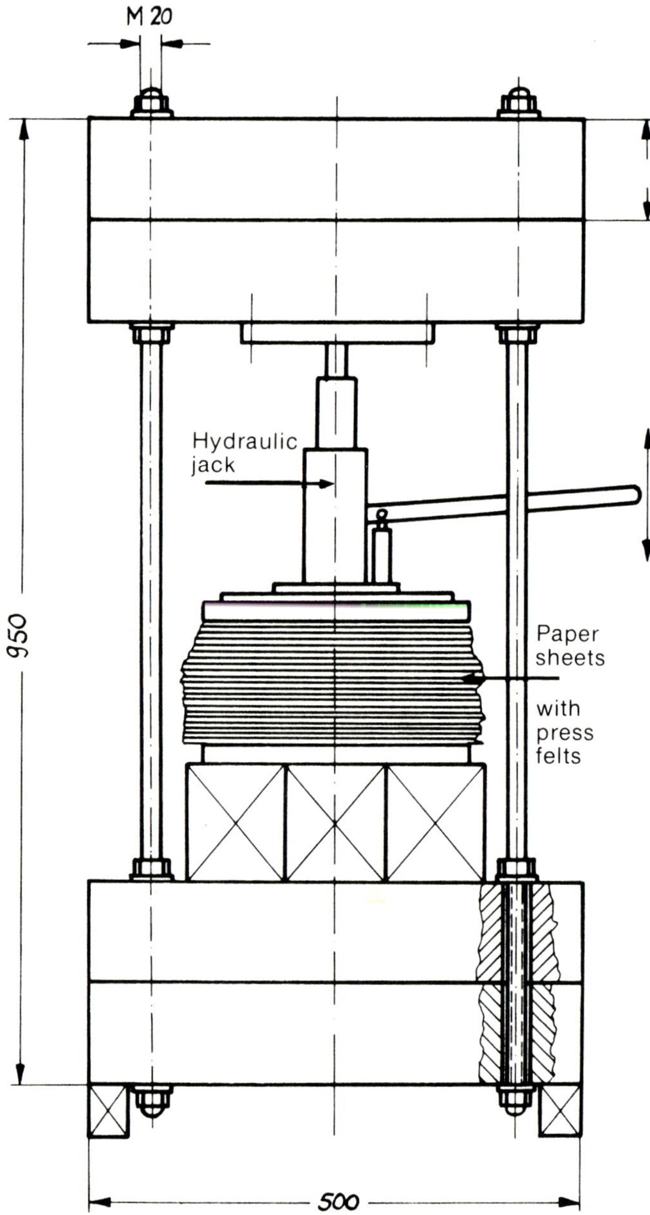

FIGURE 5: Construction schematic of a paper press incorporating a hydraulic jack to develop the pressing force. Sketch at lower left shows the hole pattern for the cross-frame members, through which the threaded tension rods pass.

Several short wooden beams are laid across the lower cross frame to form the press bed (see Figure 6), on which one or more press plattens are then placed. The paper sheets and their press felts are then placed on top of the platten; another platten is placed on top of entire stack.

The hydraulic jack is then placed (Figure 5) between the top platten and the base of the upper cross-frame. Using a jack with a load rating of 1.5 to 2 tons, one can generate a surface pressure of 1.65 to 2.2 kp/cm^2 on typical A4-size sheet.

Instructions for building a press

The cross-frame members should be made from good, unplaned spruce beams with a 10×12 cm section. If possible, the pieces should be free from knots and cracks.

The pieces should be sawn to the correct length with good square ends, and then the holes for the tension rods should be carefully drilled (Figure 5). The wider face (12 cm) of the beams forms the pressing surfaces of the upper and lower cross frames (Figure 6).

When drilling the holes (22 cm diameter) it is best to use a good, sharp, spiral wood drill bit to get a clean hole; a drill press is also recommended.

Since deep holes often run a bit off the vertical, it is best to drill one cross frame piece first, then use it as a template for locating the holes on the other pieces. The centerpoints for the holes can be scored using the drill bit itself. The cross frame pieces should be numbered to identify their positions. This method will help assure that the tension rods pass through the cross frames cleanly.

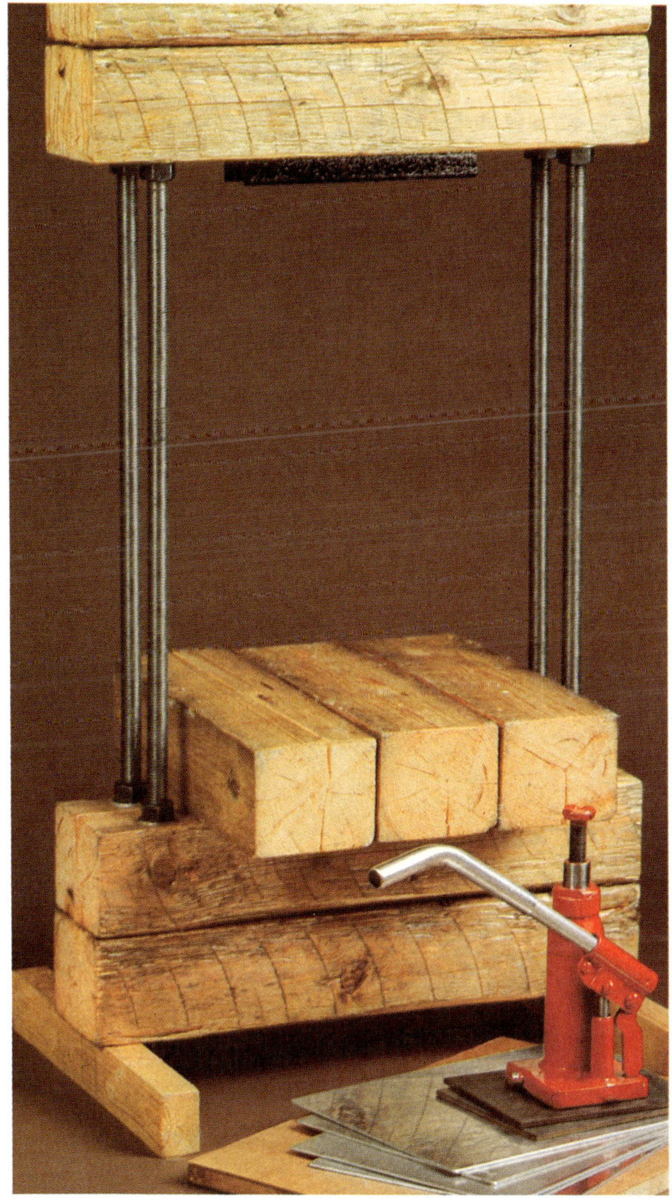

FIGURE 6: The complete paper press consists of wooden upper and lower cross-frames, wooden feet on the lower cross-frame, and a steel reinforcing channel on the underside of the top cross-frame. The press bed is made up of three wooden beams fastened to the lower cross-frame.

The four cross-frame members are then assembled with the four threaded tension rods. The two supporting feet should now be fastened to the lower cross-frame member with screws. (Figure 6 shows the position of the feet.)

Now the tension rods are unscrewed from the cross frames and the support beams can be attached to the top side of the lower cross-frame assembly. The support beams, 35 cm long, are cut from the same size wood (10×12 cm cross section) as the cross-frame pieces. The wider face (12 cm wide) is placed facing upwards. These support beams are fastened with flathead screws (5×150 mm) – two 5-mm-diameter holes are drilled through each piece.

Prior to final assembly, the wooden pieces should be sanded with rough sandpaper to remove sharp corners and splinters. The wood should then be treated with a primer oil, and the ends of the beams should be dipped in the oil to saturate them. (About one liter of wood primer oil is needed.) A generous coat of clear wood oil should then be applied, particularly to the open ends of the beams to prevent press water from entering.

To protect the bottom side of the upper cross frame from the ram of the hydraulic jack, a length of steel "U" channel is attached (see Figure 9). Two 5-mm screws are sufficient for this.

Now assemble the press. Be sure to use washers under each nut on the tension rods.

After assembly, lay a few extra pieces of wood on the press bed to simulate the height of a stack of paper sheets. Put the jack in place and pump it up. While the press is under load, carefully tighten all the nuts on the tension rods. Release the pressure, remove the jack: your press is now ready for papermaking.

When pressing paper, it is a good idea to lay a plastic sheet over the press bed and lower cross frame to help the press water to run off and keep unnecessary water off the wood.

FIGURE 7: Mounting the upper cross frames on the tension rods. Note the "U" channel fastened underneath the cross frame to protect the wood from the hydraulic jack.

FIGURE 8: Tighten the nuts on the tension rods carefully. Use a washer under each nut.

Table 3: Materials for paper press

Part No.	No. of Pieces	Description	Dimensions
1	4	Beams	12×10×50 cm
2	3	Beams	12×10×35 cm
3	2	Beams	5×5×35 cm
4	4	Rods, threaded	M20 mm × 1 m
5	16	Hex-head nuts	M20
6	16	Flat washers	for M20 nuts
7	2	Plywood boards	36×37×2 cm
8	4	Metal plattens	32×37×0.3 cm or thicker
9	1	Hydraulic jack	1.5–2.0 ton
10	1	U-channel, steel	4×4×20 cm
11	4	Woodscrews, rdhd	80×4 mm diameter
12	6	Woodscrews, flthd	150×5 mm diameter
13	2	Woodscrews, rdhd	30×3 mm diameter

FIGURE 9: The assembled upper cross frame, showing the "U" channel.

Chapter IV

MODERN PAPERMAKING

The first machine papermaking began in Europe at the start of the 19th century, about 500 years after the knowledge of this craft first arrived in Europe at the end of its long journey from China.

Not long after machine papermaking began, the supply of old rags – the main raw material for papermaking – became insufficient. The growing demand for more paper made the search for a substitute fiber source critical. Many trials over the years finally led to the use of wood as the paper industry's primary raw material today.

After over 150 years of continuous development, wood remains by far the dominant raw material for papermaking in almost all regions of the world. Although many other plant fibers are used – such as bamboo, sugar cane bagasse, rice and wheat straw, reeds and sisal hemp – wood is generally accepted as the most practical and preferred fiber for good papermaking.

In wood, the paper industry enjoys a natural, renewable raw material – an advantage enjoyed by few industries. This advantage grew after the energy crises of the 1970s, as the energy value of wood was realized.

Today, the paper industry has mastered the use of wood wastes for energy, and through this has been able to cut its use of fossil fuels (oil, gas, coal) dramatically. The typical pulp and paper mill today, after carefully extracting all the useful fibers in its wood, burns all remaining waste products, such as bark and sawdust, in specially designed boilers to generate energy.

The paper industry today is to a great degree independent of purchased fuels.

Wood, the renewable resource both as fiber raw material and as fuel, is considered by many to be the paper industry's greatest asset, and one which assures its secure future.

Many species of wood used today

Pine and spruce from the northern temperate climates of central Europe and the Nordic countries and North America were the first wood species to be heavily used for papermaking. Even today, they are preferred species for many kinds of paper due to their excellent fiber properties.

However, in recent years many other species of wood have been developed for papermaking, and are finding increased application. A good example is the northern birch, once considered a weed tree and now heavily used to make writing paper. The same is true of eucalyptus, which has been successfully bred as a fast-growing papermaking raw material in southern Europe, Brazil and in its native Australia. Tropical pines have also entered the industry's raw material picture, and they are now planted for papermaking in Central and South America, Africa, Asia and the southern United States.

Today's pulp and paper technology

The brief review of modern pulp and papermaking technology given here provides a valuable insight into the world of paper in which we all live. For a more complete understanding of the field, there are many excellent books available, many of which require little or no technical backgroung.

Additional information on further educational materials can usually be obtained from your country's national paper industry association.

Chemical pulp and chemical recovery

Chemical pulp is so called because it is produced by chemically treating the wood (or other fiber) to remove the lignin, which is the binding material which holds the wood fibers together. Dissolving the lignin frees the wood fibers, which become pulp.

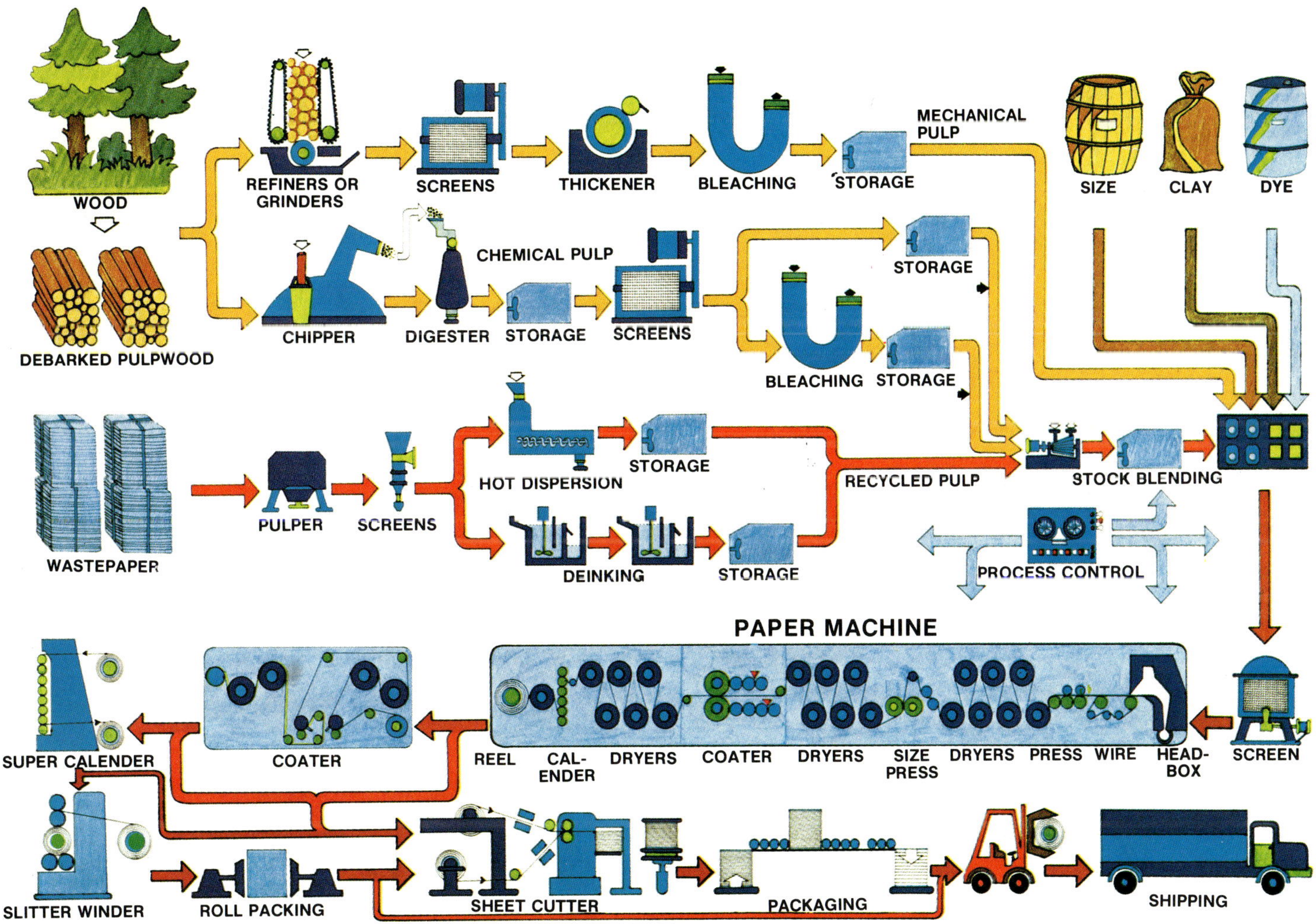

FIGURE 1: This schematic diagram explains the modern papermaking process from the forest to the finished product.

Most of the wood used, called pulpwood, is actually small-diameter wood too small for use as sawn timber. It is often from forest thinning operations, in which the weaker trees are cut away so that the stronger trees have more room and light to grow.

The pulpwood is cut into small, thin chips. To this supply, the pulp mill often adds waste woodchips from nearby sawmills or other wood processing plants.

The chips are then fed into a large pressure vessel known as a digester, into which a chemical solution is added. The digester is then heated with steam and pressurized. Under high temperature and pressure the woodchips are "cooked" – the chemicals dissolve the lignin, freeing the cellulose fibers.

Modern pulp digesters can be as much as 80 or 90 meters (250 to 300 feet) high, and a single continuous digester can produce over 1,000 tons of chemical pulp in a day.

After digesting, the resulting cellulose is washed to free it from the dissolved lignin and the cooking chemicals. Subsequently, the pulp is bleached to high whiteness.

Typical bleaching chemicals include chlorine, oxygen and hydrogen peroxide. After bleaching, the pulp is carefully cleaned to remove any dirt or non-fiber particles.

If this pulp is to be used on site to make paper, it is stored in water suspension until ready to be used on the paper or paperboard machine.

If, instead, the pulp is to be transported to another location, or sold on the open market, it is dried – usually on a modified type of paper machine – then cut into sheets and baled for transport.

Chemical recovery: The solution of lignin and cooking chemicals which is washed out of the pulp after the digester is sent to a recovery system. Essentially the liquor, as it is called, is fired into a boiler. The lignin burns away, releasing energy, while the chemical components are recovered as salts to be recycled and made into fresh cooking chemicals. Virtually nothing is lost.

Groundwood and mechanical pulps

Groundwood is made – as the name suggests – by grinding wood against a rotaiting grinding stone. Water is added during the grinding process. The resulting groundwood contains short and long fibers, splinters and fine particles called simply "fines".

This mixture must first be sorted – the splinters are rejected, of course, and returned for more treatment. Then the resulting pulp is often bleached to improve its brightness. It is, however, not bleached to the same high brightness as chemical pulps.

But groundwood is only one of several types of mechanical pulps – so called because they are made by mechanical rather than chemical methods. The other mechanical pulps are generally known as "refiner pulps".

Refiner pulps are made from woodchips, not from pulpwood logs as is the case with groundwood. In the refiner process, woodchips are ground or "refined" between rotating metal discs. Refiners, as the machines are called, are often very large, and are driven by very powerful motors – as much as 10,000 horsepower in some cases.

Some types of refiner pulps are thermomechanical pulp (made under pressure at higher temperatures), refiner mechanical pulp and chemi-mechanical pulp (in which some chemicals are added to chips prior to the refiners).

Groundwood and most other mechanical pulps are used directly at the papermills to make paper or board. Only a small amount is sold as market pulp.

Spruce makes the best groundwood, and coniferous species in general make the best mechanical pulps or all types.

FIGURE 2: The forming section or "wet end" of a modern paper machine. This machine produces a web of paper over 6 meters wide at speeds over 1.000 meters/minute.

VALMET

The yield of mechanical pulps – the ratio of finished pulp to raw wood input – is generally above 90 %. This is almost double the level of bleached chemical pulps, which commonly have yields of around 50 %.

Uses of chemical and mechanical pulps

Since chemical pulp consists mainly of whole, undamaged fibers (about 4 mm long), it can provide the paper with good strength properties. Naturally, depending on the pulp's brightness and other properties, it can lend whiteness, opacity and other characteristics to the paper made from it.

High quality grades of paper – including writing paper, document and security papers, and many special papers – are made from chemical pulp, sizing and various additives.

Many mass market grades of paper use significant quantities of groundwood and other mechanical pulps. This is particularly true of printing papers such as newsprint, rotogravure and offset printing papers, including both coated and uncoated magazine papers. These grades are the largest consumers of mechanical pulps. Groundwood and other mechanical pulps improve the printability of paper, and are thus of such importance in publication grades.

Wastepaper and recycled fiber

Large quantities of wastepaper are used today in the production of many grades of paperboard, used to make boxes and cartons. Wastepaper is also often one of the components used in making newsprint.

Recycling of another paper is an "environmentally friendly" product.

There are many ways in which wastepaper can be recycled, and the methods employed depend on both the type of wastepaper being recycled (old boxes, newspapers, magazines) and the product to be made.

Deinking is a process in which newspaper and magazines can be recycled for re-use. As the name suggest, the process removes the ink from the paper fibers, leaving a clean pulp which can be re-used in a variety of papers – from newsprint and other printing papers to hygienic papers and board grades.

FIGURE 3: A modern pressurized grinder makes high-quality pressurized groundwood pulp

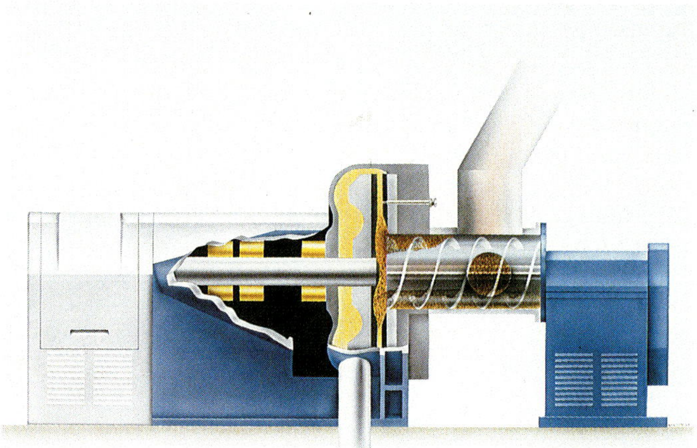

FIGURE 4: Thermomechanical (TMP) and Chemi-thermomechanical (CTMP) pulps are produced in this refiner.

The modern paper machine

Each grade of paper and board requires a very specific set of components – chemical and mechanical pulps, recycled fiber, sizing and additeves – and each component must be prepared carefully, according the specifications of the paper or board to be made (just as in hand papermaking).

When all the components are correctly prepared – all dissolved in water – they are mixed in their required proportions and this "stock" is then given a final cleaning to remove any last remaining dirt or unwanted foreign particles. It is now ready for the paper or board machine.

The paper machine has a continuous forming wire moving at high speed. Very dilute pulp stock is distributed onto the moving wire very evenly by the so called headbox. The stock is only about 1 % fiber and 99 % water when it leaves the headbox. It is rapidly dewatered along the length of the moving wire, as the wire passes over foils and suction boxes.

At the end of the wire section, the paper is couched off the wire by a couch felt – also moving at high speed and using suction to pull the paper off the wire – and is fed into a series of presses. In these presses the paper "web" is squeezed between large rolls, loaded with tons of force. The paper web is at all times – just as in hand papermaking – protected by a press felts during the pressing process.

Finally, the semi-dried web leaves the presses and enters the dryer section. At this stage, 40 to 45 % of the original water has been removed from the web. The dryer section of a paper machine consists of a series of cast iron cylinders, usually about 1.5 meters in diameter. There can be as many as 40 or 50 of these cylinders in series.

FIGURE 5: At the end of the paper machine, the paper is wound up onto large reels, which are taken on to further cutting and finishing operations.

The paper web is threaded over the rotating dryer cylinders, which are heated internally with steam. The heat carefully dries the paper.

Located midway in the dryer section is often a size press to apply sizing to the paper. This is normally done with writing papers.

Finally, at the end of the machine the paper is wound up on a large reel. As the paper roll approaches its maximum diameter, it is removed from the machine and a new roll is started.

Then follow the finishing operations. The paper may now be coated to improve its printability. Or perhaps it will be calendered to improve its smoothness and make it glossy.

The large rolls of paper are slit and trimmed in "rewinders" into smaller reels. Some of these are packaged for shipment. Other reels are fed into machines which cut the paper into large and small-format sheets – writing paper, copy paper, printing papers – which are then also carefully wrapped and packed for shipment to consumers.

Modern paper machines are among the largest machines built by any industry. The largest paper machines in the world are over 10 meters wide, 100 meters long and weigh thousands of tons. The fastest machines operate at speeds well above 1,000 meters/minute. Newsprint has been produced at speeds in excess of 1,200 meters/minute. Tissue paper has been made at around 2,000 meters/minute, which means a web of paper 120 kilometers long produced each hour.

Paper's aging characteristics

Favorable climatic conditions have helped preserve certain examples of very early paper and papyrus over many centuries. Rag containing papers also exhibit excellent resistance to aging.

Unfortunately, it has been learned that papers manufactured much more recently do not have such good resistance. Many valuable collections of books are threatened by deterioration, and people are trying to salvage what they can, often at considerable expense. What is the cause of this?

With the introduction of modern wood fibers in papermaking, certain chemical additives were employed which lead eventually to apparent deterioration of the fiber network. Papers containing groundwood pulp have the shortest resistance to aging: thus newspaper archives are threatened. Naturally, deterioration is also accelerated by environmental factors such as too humid storage conditions and air pollution.

Book and paper restorers have in the meantime become much sought after specialists. And, it is one of major undertakings of our times to preserve the existing cultural heritage which has come down to us in the form of texts and pictures on paper, so that future generations may enjoy it.

In recent years, paper grades have been developed, made from chemical woodpulp, that have virtually the same resistance to aging as rag-containing papers. These papers are commonly used for special documents and pictures.

Paper formats

People began to establish uniform paper formats quite long ago. The desire for uniformity had, of course, practical reasons. In the old libraries, the bound books and documents had to have the same size in order to be neatly arranged on shelves. With the discovery of the art of printing came a further incentive for standard paper formats; namely the size of printing press and subsequent modern printing and converting equipment. Finally, modern business practices have a major influence on establising paper formats: without today's standard formats the modern postal and banking systems would be hardly imaginable.

In many European countries, the so-called DIN standards are used. (DIN stands for Deutsche Industrie Normen, or German Industrial Standard.) Each DIN format is designated by a capital letter A, B, C or D, followed by a number. Each letter designates a format series, while the number indicates how often the largest size of that series ("0") must be folded – or subdivided – to obtain the designated format. For example, the common European format for letter and writing paper is DIN A 4; it is obtained by folding a sheet of A 0 four times. The "A" series is commonly used for printing and writing papers.

For "dependent" formats, such as envelopes, file folders and so on the format series B and C are used.

The DIN A 0 format sheet dimensions of 841 mm by 1189 mm have a surface area of exactly 1 square meter. Thus the weight of an A 0 sheet is exactly equal to the basis weight of the paper in g/m2. Incidentally, the height to width ratio of all DIN format sheets is expressed by the formula 2:1=1.414

Thus, 1189:841=1.414.

In a DIN format series the next smaller size is always formed by folding the sheet in the middle of its long (or height) dimension. Thus, by folding an A 3 sheet in half, two A 4 sheets result. Since the relationship of the sides is a constant equal to 2, all DIN formats are geometrically similar.

In Europe, the DIN format system is used in Belgium, the Netherlands, Germany, Norway, Sweden, Finland, Austria, the USSR, Switzerland, Czechoslovakia, Romania, Poland and Hungary. In Asia, both Japan and China also use this system.

In the United Kingdom and in the United States and Canada, other formats are used, as shown in the second table below.

Table 1: DIN format sizes

(in millimeters)

Designation	Series A	Series B	Series C
A 0	841×1189	1000×1414	917×1297
A 1	594× 841	707×1000	648× 917
A 2	420× 594	500× 707	458× 648
A 3	297× 420	353× 500	324× 458
A 4*	210× 297*	250× 353	229× 324
A 5	148× 210	176× 250	162× 229
A 6	105× 148	125× 176	114× 162
A 7	74× 105	88× 125	81× 114

Note: The series continues in this fashion to A 10. * denotes the standard format for letter paper.

Table 2: English/American format sizes

(Dimensions in inches)

Writing papers	Printing Papers
Foolscap 13–1/4×16–1/2	Large foolscap 13–1/2×17
Small post 14–1/2×18–1/2	Crown 15×20
Small Demy 15–1/2×20	Large post 16–1/2×21
Large post 16–1/2×21	Demy 17–1/2×22–1/2
Small medium 17–1/2×22	Medium 18×23
Medium 18×23	Royal 20×25
Small Royal 19×24	Large Royal 20×27
Super Royal 19×27	Super Royal 20–1/2×27–1/2
Imperial 22×30	Imperial 22×30
	Double Crown 20×30
	Double Demy 22–1/2×35
	Quad Crown 30×40
	Quad Demy 35×45

Biographies

Ralf Weidenmüller, the author of this book, began making paper by hand in 1954. An engineer, who spent 14 years working in the paper industry, he carried out many trials and tests and developed his own hand papermaking equipment. He has, over the years, given numerous demonstrations of the hand papermaking technique to schools and adult groups all over Germany.

Ralf Weidenmüller is the general manager of Zellcheming, the Association of German Pulp and Paper Chemists and Engineers, in Darmstadt, and is at the same time the Editor-in-Chief of the German paper industry magazine "Das Papier". He is also General Secretary of Eucepa, the European liaison committee of national pulp and paper industry technical associations.

John Kalish, the translator of this book, is an American, born near Philadelphia, not far from the site where Wilhelm Rittenhausen first made paper in the United States. He graduated from Lehigh University in Pennsylvania with degrees in engineering and economics. He moved to Germany and worked in Göttingen in 1967. In 1968 he moved to Brussels and began working as an editor at PPI -- Pulp & Paper International, the paper industry magazine of Miller Freeman Publications based in Brussels. Since 1983 he has been Editor and Publisher of PPI.

In over 15 years of reporting on the paper industry, he has visited papermaking operations on four continents. He learned of this book through his personal acquaintance with the author, whom he thanks for having offered him this opportunity to learn the art of hand papermaking.